Best Wishes

Shrimp

The Endless Quest for Pink Gold

Jack Rudloe and Anne Rudloe

Vice President, Publisher: Tim Moore
Associate Publisher and Director of Marketing: Amy Neidlinger
Editorial Assistant: Pamela Boland
Development Editor: Kirk Jensen
Operations Manager: Gina Kanouse
Digital Marketing Manager: Julie Phifer
Publicity Manager: Laura Czaja
Assistant Marketing Manager: Megan Colvin
Cover Photo: Dave Barfield
Cover Designer: Chuti Prasertsith
Managing Editor: Kristy Hart
Project Editors: Julie Anderson and Jovana San Nicolas-Shirley
Copy Editor: Gayle Johnson
Proofreader: William Woods Publishing Services
Indexer: Lisa Stumpf
Senior Compositor: Gloria Schurick
Manufacturing Buyer: Dan Uhrig

Publishing as FT Press
Upper Saddle River, New Jersey 07458

FT Press offers excellent discounts on this book when ordered in quantity for bulk purchases or special sales. For more information, please contact U.S. Corporate and Government Sales, 1-800-382-3419, corpsales@pearsontechgroup.com. For sales outside the U.S., please contact International Sales at international@pearson.com.

Printed in the United States of America

First Printing December 2009

ISBN-10: 0-13-700972-0
ISBN-13: 978-0-13-700972-5

Pearson Education LTD.
Pearson Education Australia PTY, Limited.
Pearson Education Singapore, Pte. Ltd.
Pearson Education North Asia, Ltd.
Pearson Education Canada, Ltd.
Pearson Educación de Mexico, S.A. de C.V.
Pearson Education—Japan
Pearson Education Malaysia, Pte. Ltd.

Library of Congress Cataloging-in-Publication Data

Rudloe, Jack.

Shrimp : the endless quest for pink gold / Jack Rudloe and Anne Rudloe.

p. cm.

ISBN 978-0-13-700972-5 (hardback : alk. paper) 1. Shrimp fisheries. 2. Shrimp culture. 3. Shrimps. I. Rudloe, Anne. II. Title.

SH380.6.R83 2010

639'.58—dc22

2009034974

We dedicate this book to Edward Keith, Glen Buffkin, Geno Litcheldello, Nick Mosconis, Joe Morgan Cruse, Fuzzy Lively, and all the other shrimpers who died at sea or passed away after a lifetime of providing everyone with delicious shrimp.

Contents

Foreword

I became acquainted with the shrimping industry four or five years ago when I was asked to do a TV commercial for the Alabama Shrimpers' Association. It was under a severe strain because of foreign pond-raised shrimp being dumped into the market. The object of the campaign was to get people to eat—and stores and restaurants to sell—Alabama, or at least Southern, wild-caught shrimp. I thought it was a worthy cause. A large seafood wholesale house I am familiar with once had nearly a hundred shrimp boats. Now what it has is big 18-wheelers at the dock unloading box after box of frozen pond-raised shrimp from Asia, which the wholesaler repackages in its own boxes with its logo on them.

I guess I was a logical choice to do the TV ad, since so many people associated the character in my book *Forrest Gump* with shrimping. In fact, I am reliably informed that in a *New York Times* survey that sought to find out why the *per capita* consumption of shrimp had skyrocketed from 1.4 pounds per person per year in the 1980s to 4.1 pounds in 2007, a significant number of people responded that it was because of the *Forrest Gump* movie. Whether this is true I don't know, but it has certainly left me with a sense of connection to shrimp and shrimpers.

And now along come the Rudloes, who, in *Shrimp*, bring an entirely new dimension to the subject, much as William Warner brought to the Chesapeake Bay blue crab in his unforgettable book *Beautiful Swimmers*.

Rich in stories of seamen and the sea, *Shrimp* tells a fascinating story not only of the little creature itself, but of all the trials and tribulations the intrepid shrimper goes through to harvest it up. As scientists and marine biologists, the Rudloes know whereof they speak, and as writers, they have the talent and grace to make it interesting.

This is one of those rare books that will do honor to any bookshelf.

Winston Groom
Author, *Forrest Gump*

1

Going Shrimping

Captain Nick Mosconis was an unshaven gruff old Greek fisherman with short gray hair. He wore frayed greasy coveralls and boots that had once been white but were now so scuffed and covered with fish scales that they had become colorless. He and his deckhand Pea Picker, a boy from Alabama, were getting ready to cast off from the fishing docks in Carrabelle, Florida at sunset. At age 18, I (Jack) had managed to get a job collecting marine specimens for a biologist at a local university. The scientist needed live batfish, a little brown leathery-skinned fish that can barely swim but walks on the bottom like a frog, and I had gone to the waterfront hoping to catch a ride on a shrimp boat to try and collect some.

As I tried to explain my need for live batfish, Captain Nick cut me off. "I don't care about your fish, or if you ride along or not. Just stay out of the way, and don't cause me no trouble."

With some trepidation, I hopped aboard his forty-five-foot trawler, the *Eurofomia*, carrying buckets and gear, and threaded my way through the clutter of nets, rusty cables, and shrimp baskets on deck. It was the worst looking, worm- and rust-eaten derelict in the boat basin. Once a schooner used by Nick's grandfather for sponge diving in Tarpon Springs until the sponge trade played out, his father used it to fish for snapper and finally Nick had it retrofitted with a diesel engine and winch for shrimping.

My worry gave way to excitement an hour later as we headed out of the Carrabelle River into the bay, the propeller churning up a wake of phosphorescent phytoplankton in the darkness. When Nick and Pea Picker hauled in the net for the first time, my life changed forever. The sheer volume and diversity of life brought up from beneath the waves shocked and overwhelmed me.

It seemed like every creature in the ocean was disgorged onto the deck, all at one time, and all in one place. Blue crabs, spider crabs, calico crabs crawled, fish flapped, while pink shrimp exploded all over the deck. Their eyes glowed like hot coals beneath the overhead lights as they jackknifed and flipped on top of the bright yellow and orange sponges.

I picked out the batfish from the pile, but Nick and Pea Picker's eyes were only for shrimp, and the large edible fish like flounder, whiting, and mackerel, which they could sell. Everything else was an impediment to get out of the way, unwanted and unloved "trash" that had to be shoveled overboard, all to the joy of a huge flock of nighttime seagulls that hovered screeching behind the boat, gobbling down the dead and dying fish as we threw them back into the ocean. The fish that hit the seafloor would be pounced upon by scavenging crabs and by the shrimp themselves. In later years this would become known as bycatch and would unleash an avalanche of environmental opposition to the shrimpers' wanton destruction of nontargeted species.

Seeing me looking at all the other strange creatures, Nick growled, "Boy, we ain't got all night. You better get a move on and cull these shrimp off, or I'll set you off on the nearest buoy! I'm gonna haul the net back again in thirty minutes, and you've still got an hour's worth of culling in front of you right now." Clearly on probation, I did my best to speed up, tossing the shrimp into the basket and raking the trash fish through the scupper hole. But the live shrimp were hard to grab, and I was repeatedly stabbed and stuck by their serrated head spines. Nick left the wheelhouse to look over the pile, and he snatched a double handful of shrimp from my discards. "Son, this ain't worth a damn," he said angrily, holding them by their whiskers and thrusting them in my face. "If you keep throwing away the shrimp like this, you ain't going back out with me again. Now cull through that again, and don't you shovel no more of my shrimp overboard, you hear?"

All through the night, the nets came up and the nets went down. Tow after tow after tow, the fishermen caught hoards of croakers, catfish, bonnethead sharks, glowing phosphorescent sea walnut jellyfish, and orange starfish, all of which would have had a marine biologist ecstatic, but which bore little interest for the fishermen. They had their own names for some creatures—names that no invertebrate biology book would ever print. Sea cucumbers were "sea pricks." "Sea pussies" were a soft squishy snail without a shell that exudes purple ink and resembles its namesake.

Our baskets filled with shrimp, and in the morning when we came into the dock, I helped Pea Picker and Nick pinch off their heads. Headless shrimp brought a better price at the fish house but lost a third of their weight. We cooked some on the boat. With their sweet taste and firm meaty texture, they were the best thing I ever ate!

Captain Nick fronted some shrimp to me to try to sell. When I got paid, I'd pay them. When word got out at the university biology department that I had come in with fresh shrimp as well as batfish, professors, graduate students, and departmental secretaries all came running. As I scooped out the pink shrimp, their eyes glowed, their mouths watered, and their wallets came out.

Shrimping soon became a way of life. In the spring and early summer, we dragged Carrabelle Bay at night for pink shrimp, or hoppers, as the fishermen called them. North Florida pink shrimp jump higher and faster than either brown or white shrimp. Reddish brown with a big purple spot on the third segment of their tails, hoppers are a subspecies of pink shrimp, much different from their clear pale pink southern cousins that live on white marl bottoms near the Dry Tortugas and Key West, Florida.

As summer advanced, we went to Apalachicola Bay and caught brownies. When fall came and cold fronts blew down from the north, white shrimp left the rivers, and we caught them just off the beaches of the barrier islands, St. George and St. Vincent. Eventually I began to send marine specimens to other universities, creating what would eventually become the Gulf Specimen Marine Laboratory, a marine biological supply and education center in Panacea on Florida's north Gulf coast.

A few years later, by sheer luck I managed to talk my way into a temporary job collecting marine specimens for the Smithsonian Institution as a participant on the UNESCO-sponsored International Indian Ocean Expedition in Madagascar. When the director of the French Center of Oceanography and Fisheries, where the American participants were staying, heard that I had worked on shrimp boats in the Gulf of Mexico, he asked me to go along on the French colonial government's first shrimping trip. Stepping aboard the French research vessel *Ambariaka* in Nosy-Be, Madagascar, I never dreamed that I would be part of history, bringing shrimping to this small impoverished nation off the southeast coast of Africa. Our plan was to trawl for shrimp near a stretch of shore where two Sakalav villages caught shrimp by seining them along the beaches and trapping them in bamboo weirs.

The *Ambariaka*, a 45-foot diesel-powered vessel, traveled all around northern Madagascar, taking measurements of salinity, dissolved oxygen, and temperature and gathering plankton samples. No shrimp boat, she was rigged for hydrographic research with a Japanese long line winch aboard, large quarters for crew and technical staff in the stern, and barely any space on the bow to handle a large heavy shrimp net.

We left the tropical paradise island of Nosy-Be at dawn and headed south, down the coast until the transparent blue waters of the coral reefs gave way to the muddy waters of the Ifasy River estuary in the Baie d'Amboro by mid-afternoon. Shrimp love estuaries, places where the rivers meet the sea, and the signs were good that we'd catch some. Shrimp love mud, and plenty of it was eroding off the deforested mountains. The forests were being cleared for firewood and farming, and with every rain, mud flowed into the streams and creeks and out into the thicket of mosquito- and crocodile-infested mangroves. A mountain chain rose up behind the mangrove swamps. We could see villages with thatched-roof houses where people barely squeezed out a living from the sea.

Although I was no expert on shrimping, I probably knew more about it than anyone else on board. We unpacked the brand-new shrimp net that the French government had sent down, spread it out on deck, and attached it to heavy wooden or metal panels called doors that would hold it open underwater. This net was different from the trawls I was used to in Florida, but a trawl is still a trawl. It's been around since the late twelfth century in one form or other, and fishermen have been trying to get it to fish right ever since. If the net is set out properly, so that the large doors don't cross and twist the net into a big knot, the doors spread apart from the pressure of the water as they are towed and hold the mouth of the net open from side to side. Floats at the top and lead weights on the bottom hold it open vertically so that the trawl skims over the bottom like a big sock, scraping the surface of mud and engulfing everything in its path. Tickler chains are dragged just in front of the net, which makes the shrimp jump out of the mud and get swooped up by the net. But if too much cable is put out, or the boat drags too slowly, the doors dig into the bottom, loading down the net with mud until it catches nothing. Or if the net is pulled too fast or is too light, it floats above the bottom and catches nothing. Getting an otter trawl to fish properly on different types of bottom is an art form that is mastered only after years of experience.

The doors needed a winch to lift them, but the *Ambariaka* had no hoisting equipment. We would have to rely on our muscles. Together we lifted the doors and set them over the high sides of the ship, trying to keep them from getting tangled. When the doors sank, we moved ahead slowly, pulling the long funnel-shaped net over the unknown and never-before-trawled bottom until the net bogged down in mud and stopped the boat cold.

Our crew hauled back the ropes with great grunts, groans, and heave-ho's. Even with the entire crew of four Comoran Islanders, four Sakalava, and three scientific staff straining and groaning, it was impossible to lift. Even if we had a winch, it's doubtful we could have pulled up the two tons of mud that hung down from the tail bag.

"Tie it off," I panted, "and let the boat wash the mud out of the net. That's the only way we'll ever get it on deck!" Dr. Pichon, the chief scientist, translated to Captain LaBarse, who nodded in approval and gave the orders. We tied down the gorged net securely and pulled it alongside the boat. As we moved up and down the shore of the estuary, clouds of mud washed out of the webbing, and the net became lighter and lighter. It was so choked with sediment, mangrove leaves, sticks, and debris that it was impossible to see any marine life in it. At last we hoisted it to the mast, tied it off, and struggled with the complicated knot that held the net closed. When it finally opened, five hundred pounds of smelly mud disgorged on the deck, wriggling with marine life.

People have caught shrimp for untold centuries in tropical estuaries around the world using cast nets on beaches or by setting seines to block creeks so that shrimp leaving the creek as the tide fell were trapped. When the shrimp moved offshore, they were out of reach and inaccessible. But now we were hauling them up from twenty feet of water, a mile or more from shore. With the trawl and the powerful diesel engine to pull it, the balance of power tipped in the direction of humans.

Even with the deck hose running to wash the rest of the mud overboard, we couldn't really see shrimp; we had to feel for them. There were crabs galore, and catfish, and croakers—the same types of estuarine animals that shrimpers caught in the Gulf of Mexico—but the shrimp catch was poor.

To catch shrimp with trawls, the nets have to be adjusted to work perfectly, neither tangling up and flipping over, nor floating too lightly above the bottom, nor floating too heavily and digging too deeply into the mud. And shrimp must be there in the first place in some appreciable quantity to be caught. Our net was plowing too deeply through the mud, and I had no idea if we were there at the right time.

"They are very happy to see the shrimp," Dr. Pichon said, watching the crew holding up our few shrimp by their antennae, chattering gleefully, and throwing them in a bucket. "Some of the boys from the village down the coast used to fish for them with seines. But we hoped there would be more. This is not so good a catch."

"It's still afternoon. Maybe these shrimp only come out at night. That happens with pink shrimp or brown shrimp in the Gulf of Mexico," I explained. "Or they may only come out on the dark of the moon, and this isn't the right time. You have to keep trying until you learn their behavior."

We set the net back out, shortened the tow rope so that it wouldn't fish so heavily, and had dinner in typical French fashion, with a white linen tablecloth, wine, bread, and wonderfully prepared fish and shrimp from the net. Nothing compares with scooping up a few pounds of wriggling shrimp from a catch, throwing them into a pot, and eating them a few minutes later. We had only enough shrimp for an appetizer, but the crab soup was delicious. Then came a big silver platter of broiled yellowfin tuna steaks that we had caught en route to the shrimping ground, served with lemon sauce. Pichon, seeing my surprise, laughed, "We live good in oceanography!" He raised his glass of wine in a toast: "Here's to good shrimping!"

A month later, we tried again, and things improved. On the last trip, the sea had been calm, but now it was turning ugly. As we towed the net in the same area, the sun sank behind the mountains, and large ominous rain clouds began blowing from the land out to sea. We were in the middle of the rainy season, and it could deluge for days on end. There was a crash of thunder. Lightning zippered across the gray skies as the *Ambariaka* strained ahead, towing the heavy nets. Then came blistering rains that beat in our faces and drove everyone into the cabin. We waited for it to stop, but it didn't.

The sea whipped into a fury, and gusts of wind-driven rain blew down on us.

The way the *Ambariaka* rode the waves made even the strongest seamen queasy. I could feel my stomach turning as the wind blew hard in our faces and the boat began to rock. "She rides the waves like a wooden bucket," Captain LaBarse said cheerfully. "When we hit even a little choppy sea, the crew bounces around like ping-pong balls. Everyone gets sick on the *Ambariaka*."

Sick or not, I finally told Pichon that the net must come up immediately, or it would be too heavy to lift. If that happened, we'd have two choices, both of which involved diving overboard in the night sea to either untie that unwieldy knot or cut open the bag and let everything loose. With big hungry sharks drawn in from the Mozambique Channel by the tantalizing fish struggling in the net, neither prospect was an option.

The moment we went through the cabin door, the *Ambariaka* lurched violently. I skidded across the deck and almost went overboard, clutching the railing with all my might. Fortunately, one of the crew caught me by the arm. We grabbed the ropes and began pulling them back, and inch by inch the net emerged from the rolling sea. Between the black webbing I could see the pink bodies of shrimp. At last we managed to swing the bag over the rail and tie it securely to the mast. The glowing eyes of endless shrimp reflected off the deck lights.

The crew tried vainly to hold down the bloated bag while Captain LaBarse tried to untie his complicated fisherman's knot, but there was no stopping it. With each lurch, the bag gathered momentum, and the fish spines and crab claws protruding from the webbing didn't make it any easier. It slammed into Pichon and me and bowled over the crew.

Captain LaBarse cursed the crew in French and Malagasy. With a yell of rage, he pushed the nearest man, who in turn crashed into another crewman. He pulled out a knife, and for a second I thought he would commit mayhem on the crew, but instead he attacked the net. Sawing savagely, he cut through the knot.

Everything poured out of the net as the webbing was lifted. In an instant the crew was waist deep in a mountain of shrimp, big jumbo shrimp, mixed in with multitudes of silvery little fish and starfish. The deck was piled so high with them that there was no room to move. Captain LaBarse and Pichon looked stunned at first, and then they began to shout and laugh. In that moment, they had both caught a bad case of shrimp fever. There is no describing the feeling of seeing your first deck full of shrimp. Most were medium to large white shrimp, Fenneropenaeus (=Penaeus) indicus, the Indian Ocean white shrimp, but there were also dozens of gorgeous black-and-white-striped tiger prawns, Penaeus monodon, that were nearly the size of small lobsters and weighed almost a pound. Never had I seen such monsters. They had such powerful tail flips, it took two hands to hold onto them. Years later these would become the mainstay of the Asian shrimp-farming industry.

The crew's eyes glowed almost as much as the shrimps' eyes did under the deck lights, and they shouted how good the shrimp would be to eat. LaBarse was right in the middle, culling like a madman, his black beard split into a wide grin. "We eat good tomorrow, no? This good fishing, yes?" LaBarse asked me excitedly. "In Gulf of Mexico, like this?" He pointed to the baskets of shrimp.

"*Oui*, sometimes. In Mississippi, they catch shrimp like this. It is a good catch, very profitable," I said as I grabbed small eels, lots of flounder, whiting, croakers, and various snails and sharks to preserve for the Smithsonian Institution.

But now we had another problem. This scientific survey vessel had turned into a commercial shrimping boat that had no way to preserve the catch. The *Ambariaka* lacked refrigeration, and there was barely enough ice on board to keep the beer cold. There was a real danger that all these beautiful shrimp would spoil.

"How you keep shrimp?" he asked.

"Ice—we keep them on ice in the fish hold," I replied, grabbing another handful of shrimp and raking another pile of bycatch behind me.

Pichon laughed, "The crew lives where our ice hold should be. Here we have problems to keep the shrimp, no?"

"This will help," I said, grabbing two handfuls of shrimp and pinching off their heads between my thumbs and forefingers. The "tail" or abdomen contains the meat. The "heads" have the whiskers, brain, digestive glands, and gonads.

"We must break the heads, or they'll turn red and spoil in this heat," I said with the officiousness of a twenty-year-old who knows everything. At home it was commonplace to pinch off the heads, usually aboard the vessel.

"But Monsieur Rudloe, that is no good," Pichon protested in shock. "You are wasting the heads! They have all the flavor! You lose the good taste when you break the head. All the—how do you say—body juice escapes, and it does not taste as good." He was right. Nothing is better than head-on shrimp steamed in beer. In Spain they're eaten no other way, but they have to be eaten fresh. The farther from the sea you are, the less likely you are to get a shrimp with its head attached.

I was convinced our precious catch would spoil, but the crew had another idea—an ancient one that has been used to preserve shrimp all over the world, perhaps since the days when man first learned about fire. They set up great pots of boiling salt water. Buckets of shrimp were dumped in and boiled until they turned a bright pink. Then the cooked shrimp were dumped out onto large wooden trays that were stored on top of the cabin so that they could drain, and more shrimp were boiled. The cooked shrimp would be dried and would last until they got to the open-air market back in Nosy-Be.

The *Ambariaka* had done it—we were the first to wet a net and find shrimp. Our catch was the first official reporting of shrimp landings. With foreign investment that began three years after our exploratory shrimp trips, local fishermen put outboard motors on their pirogues and canoes, the government sold licenses, and foreign governments brought in large trawlers with freezers that produced millions of pounds of shrimp to be sold to Japan, Spain, and other countries. Today Madagascar shrimp have become a trade name in France. One of the poorest nations in the world, where the average wage is still two dollars a day, the Malagasy Republic became Africa's leading shrimp producer as shrimp became its third-largest export. Today millions of pounds of shrimp are harvested off the same coast, and some of those beautiful wild shorelines and mangrove swamps I saw nearly fifty years ago have been converted into shrimp farm ponds.

2

About Shrimp

Shrimp are abundant, easy to catch, and delicious. Whether they're fried, steamed, boiled in beer, or served with sauces and dips, people love the sweet succulent flavor of shrimp. This is a book about shrimp, people who love to eat shrimp, fishermen who roam the seas catching them, and farmers who grow them in ponds. And it's a book about a conflict that is as old as humanity itself, hunter-gatherers versus agriculture—or, in this case, "aquaculture." Farmers have provided such a huge global supply of pond-grown cheap shrimp that low prices have nearly destroyed the fishermen.

All of this human behavior and activity swirl around small but enormously abundant crustaceans with long whiskers, bulbous eyeballs on stalks, ten legs, and a delicious tail. They can flip themselves off the seafloor and then dive into the mud to escape predators. Shrimp are laterally compressed crustaceans (flattened from side to side) with a streamlined shape for swimming. They have slender delicate legs for perching briefly on the bottom rather than the stronger walking legs of crabs and lobsters. They swim forward with paddle-like appendages called pleopods on the underside of the abdomen. Backward swimming is accomplished by tail flipping of their uropods, the flat plates on the tail that enable propulsion (see Figure 2.1). Depending on which expert is asked, there are approximately four thousand species of shrimp. Different species occur from the tropics to the arctic, from the edge of the sea to the ocean depths and in freshwater lakes and streams. All shrimp are essential links in marine food chains. Several shrimp species support multibillion-dollar commercial and recreational fisheries.

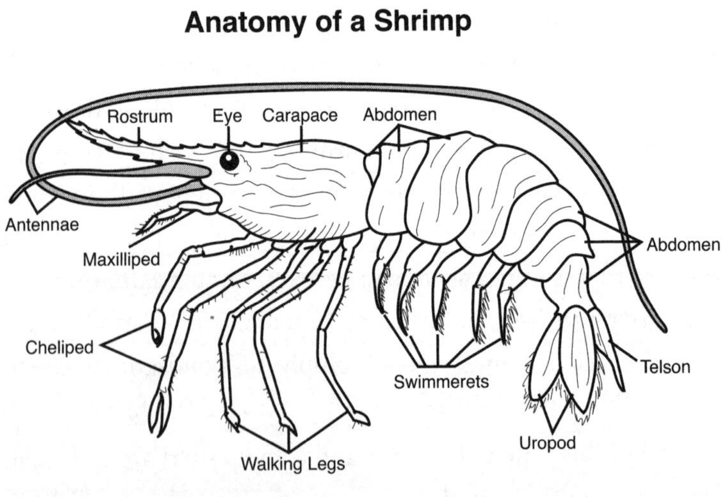

Figure 2.1 Anatomy of a shrimp

Courtesy of the Louisiana Sea Grant College Program

Approximately four hundred of those species, collectively called the penaeids, release their eggs into the ocean. These hatch as very undeveloped nauplius larvae, the first and simplest developmental stage of all crustacean species. A nauplius-stage shrimp is little more than a swimming head with the thorax or middle parts of the body and abdomen still absent. Among the twenty-eight hundred species of mostly tiny caridean shrimp, females carry their eggs on their abdomens until they pass through the nonfeeding nauplius stage. Then the eggs hatch as a more developed larval stage, which has more segments and appendages. Most carideans are tiny, inconspicuous, and known only to marine biologists, but a few species are large enough to eat and support fisheries.

In both the penaeids and carideans, the newly hatched baby shrimp are barely visible to the eye, look nothing like the adult shrimp they will become. They live independently in the plankton as they grow, molt, and develop through a long series of larval stages. When they become fully formed juvenile shrimp, they leave the plankton and take up life on the seafloor. Figure 2.2 shows the life cycle of the shrimp.

The approximately eleven hundred species of opossum shrimp or mysids are small shrimp like crustaceans, only distantly related to the penaeid and caridean shrimp. The females have brood pouches in which the developing young complete all the larval stages prior to hatching as small but fully developed mysids. Another group, the stenopodids, include only about twenty species of shrimp, but many of them, such as the banded coral shrimp, are beautifully colored and are popular in home aquariums. They have large claws and are more closely related to crabs and lobsters than to the penaeid or caridean shrimp.

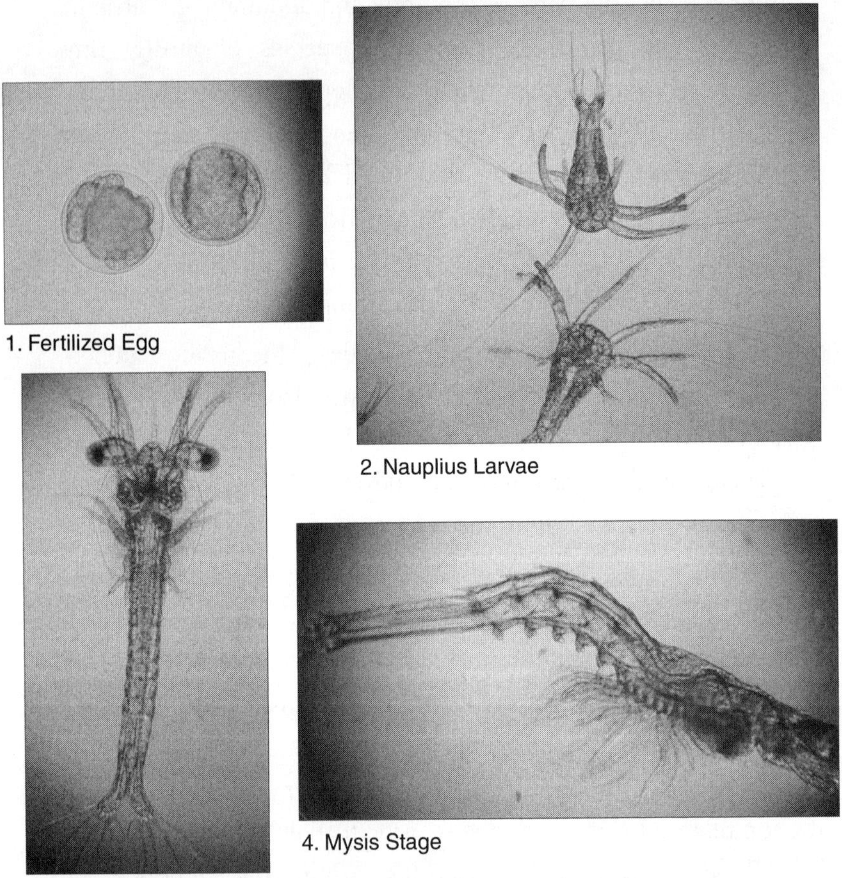

1. Fertilized Egg

2. Nauplius Larvae

3. Protozoea Larvae

4. Mysis Stage

Figure 2.2 Life cycle of the shrimp

Courtesy of American Shrimp, LLC

And it gets even more complicated!

"So many things are called shrimp that it's confusing," chuckled Greg Jensen, an expert on crustaceans at the University of Washington, when we asked him to define the word *shrimp*. "Seed shrimp are ostracods; fairy shrimp, clam shrimp, brine shrimp, and tadpole shrimp are brachiopods; and skeleton shrimp are amphipods. Then there are mud and ghost shrimp, which are yet another group, and mantis shrimp or stomatopods. As I see it, 'shrimp' seems to be almost any crustacean that isn't a lobster, barnacle, or crab."

The word *shrimpe* originated in Britain. It means something very small, although there is nothing small about the impact of shrimp on the world. The idea of smallness no doubt came from people catching the little gray shrimp, *Crangon crangon*, in European waters. "Prawn," on the other hand, is a term usually reserved for the largest species, like *Pandalus eos* and *P. playceros*, the holy grail of shrimp.

Of all species of shrimp known to science, only a few are eaten by man. Most of those are large penaeids of several different species that occur in huge numbers. They may be sun-dried, canned, iced, or frozen, or even freeze-dried for astronauts and backpackers. Everywhere in the world, when edible shrimp show up, the human feeding frenzy begins. Outbreaks of shrimp fever occur aboard trawlers working at sea, among people in Africa dragging them ashore with beach seines, and in recreational fishermen from South Carolina to British Columbia.

In the Pacific Northwest, the six-inch-long cold-water pinks, *Pandalus jordani*, and spot prawn, *Pandalus playceros*, are commercially trapped or trawled from Oregon to Alaska, as are the closely related *Pandalus borealis*, caught off the coast of Maine. In Europe there is a food market for the little gray caridean, *Crangon crangon*, and in Asian waters fishermen set nets to catch tons of thumbnail-sized grass shrimp, *Palemonetes*.

All of those are minor fisheries compared with penaeid shrimp. Wild and farmed, penaeid shrimp are a fifty-billion-dollar-a-year industry, which makes them the most valuable creature in the sea. The vast majority of wild-caught shrimp in the United States are three closely related species known in the U.S. as pink *Farfantepenaeus (=Penaeus) duorarum*, brown *Farfantepenaeus (=Penaeus) aztecus*, and white *Litopenaeus (=Penaeus) setiferus* shrimp, from the estuaries, bays, and sounds of the South Atlantic and Gulf of Mexico. These shrimp are so similar that only a specialist can tell the difference. The commercial pinks and browns are "grooved shrimp" with a slit along the back of their last segment, in front of their tail fins. If you can insert your fingernail into the groove, most likely it's a pink. If the slit is too narrow, it's a brown, and if it has no groove at all, it's a white shrimp, which usually has a beautiful green tail fin.

There is a shrimp for every habitat. Some species seek out sea grass meadows; some dwell in soft mud, coarse sand, and shells. Shrimp are harvested wherever rivers flow to the sea, sinuously winding their way down to the bay, and slowly blending fresh water with salt within bays, estuaries, and barrier islands. Sea bobs in the genus *Xiphopenaeus,* sugar shrimp *Trachypenaeus constrictus,* rock shrimp *Sicyonia brevirostis,* and royal reds *Pleoticus (=Hymenopenaeus) robustus* are also landed in small numbers. Royal reds are found between one and three hundred fathoms by a few large commercial vessels with big winches. Miles of cable drag them up from the frigid deep waters at the edge of the continental shelf in an area that snapper fishermen call "the end of the Earth."

The scientific names of shrimp not only confuse, but they also put anyone who isn't an ardent invertebrate zoologist to sleep. Yet scientists rely on these Latin names to eliminate confusion since common names for the same species often vary from region to region and language to language. Scientific names change only because zoologists working in different places may decide that the original classification didn't accurately reflect who is related to whom biologically. Two different biologists might independently describe the same species, resulting in two different names for what later proves to be the same animal. Or the same name might be used for animals that are later discovered to actually be two different species. Then a decision has to be made as to what the beast will be called, using an elaborate set of rules based on when the species were first discovered and by whom.

Carl Linnaeus, the father of the system for classifying living species, first named the white shrimp from North America *Cancer setiferus* in 1767. When Harvard naturalist Louis Agassiz examined the white shrimp in 1849, he reclassified it as *Penaeus setiferus (Linnaeus)*. Scholars in those early days of biology were schooled in the classics. Penaeus was the Greek river god who turned his daughter, Daphne, into a tree because Apollo was after her with dishonorable intentions. The magical transformation didn't stop the great sun god from trying to make love to the laurel, but that's another story.

In due course the closely related pink and brown shrimp of the southeastern U.S. became *Penaeus duorarum* and *Penaeus aztecus*. Other species of *Penaeus* were later described from other parts of the world. All was well until 1997, when Perez Farfante and Brian Kensley published a monograph of the prawns and shrimps of the world. After examining morphological differences in different species of the genus *Penaeus*, they decided that the most commercially important invertebrates in the world were misclassified, and they instituted sweeping changes. The genus of all the warm-water pink, brown, and white shrimp should no longer bear the name *Penaeus*, but instead should be *Litopenaeus* for white shrimp and *Farfantepenaeus* for brown and pink shrimp. Other species became *Marsupenaeus* and *Fenneropenaeus* The delicious deep-water royal red shrimp that was once called *Hymenopenaus* got a new name, *Pleoticus*. With that monograph, twenty-seven well-established scientific names for twenty-seven species of penaeid shrimp bit the dust, to the annoyance of many in the aquaculture industries. Only the giant tiger prawn, *Penaeus monodon*, was allowed to keep its original name.

Since then many scientific research papers have been published, dutifully using the new names. The change is now in dispute, however. A shrimp expert in Thailand, where the world's biggest shrimp growers live, insisted that no DNA evidence justifies the renaming of all those species. Also, a loophole in the rules for bestowing scientific names allows the retention of older, well-established names. The scholarly battle rages on as different scientific authors choose to use either the new or old names in their technical publications. In the most recently published research papers, the new names are winning by a margin of 2 to 1, so we've grudgingly adapted the new names in this book. This practice may yet be reversed in the future as more taxonomic research is done.

By whatever name they are called, shrimp have been cooked and eaten for thousands of years. "There have always been customers for shrimp ready to fall upon them whenever and where they could be discovered," wrote M. Gavius Apicius, a Roman author of cookbooks in the first century A.D. One of the first recorded foodies, he cooked at his own banquets. It's said that when he heard that there were large delicious shrimp to be had in Libya, he traveled to Africa to eat them on the spot, but he was so disappointed by the size of what he saw that he sailed home without ever going ashore.

Whether they are the tiny caridean grass shrimp of the genus *Palemonetes* or the gray *Crangon* shrimp of the North Sea and China, the cold-water pinks of the North Pacific and North Atlantic, the warm-water penaeid shrimp, or the farmed tiger prawns and big freshwater *Macrobrachium* shrimp of the tropics, they're all good to eat. There are no poisonous shrimp, although some unfortunate people are highly allergic to them. Although we've never tasted the blind cave shrimp, nor have we feasted on the bizarre eyeless shrimps that live in the hydrothermal vents at the bottoms of the deepest oceans, we're sure they are delicious. We know of no one who has eaten the semi-terrestrial mangrove burrowing shrimp of New Guinea, but a shrimp is a shrimp, and it might be worth a try.

Today, farmed shrimp outnumber wild-caught shrimp in the marketplace. Before the 1990s, approximately 70 percent of shrimp consumed worldwide were caught at sea. By 2000, 50% were farmed. By 2009, 70 percent had never seen the ocean but grew up in ponds in China, Thailand, Vietnam, India, Taiwan, Ecuador, Mexico, and other Asian and Latin American countries, as well as Australia. Black-and-white tiger prawns from Pacific mangroves, *Penaeus monodon*, are farmed in Asia. Giant freshwater caridean shrimp *Macrobrachium rosenbergii* and South American white shrimp *Litopenaeus (=Penaeus) vannamei* are being raised in greenhouses in Colorado, Kansas, Alabama, and Maryland, as well as ponds and impoundments across Central and South America.

Gourmets, epicureans, and seafood purists insist that wild-caught shrimp fresh from the ocean are superior to shrimp that are raised in ponds and fed artificial feed. We prefer a wild shrimp that has spent its life roaming the oceans, eating a wide variety of foods, but we have eaten farm-reared *vannamei* that were excellent. In one shrimp taste test in Seattle a few years ago, tasters preferred wild-caught *vannamei* over wild whites and browns; farmed *vannamei* came in well down the list. Nevertheless, farmers claim their product is more uniform and has a better quality and consistency than wild shrimp that may have been sitting on ice in the hold of a shrimp boat for a week before coming to the dock. Farmed shrimp are seined out of the ponds and go immediately into the freezer, often still kicking.

The fresher the shrimp, the less flavor they have. Slightly aged shrimp taste “shrimpy,” just as hunters hang game for a few days to improve the taste. Processors produce peeled, gutless, beheaded, and breaded popcorn shrimp by letting them age until the meat separates from the shells so that they can pass easily through the shelling machines. Peeling fresh shrimp by hand is a tedious process. Some people practice shrimp peeling meditation by simply switching off their brains, putting down all their opinions, and just doing it. Or it is possible to boil them whole and let everybody peel their own at the dinner table.

The habitat where wild shrimp are caught affects their taste. White shrimp from the Atlantic are sweeter and firmer than whites from the Gulf of Mexico. The little hard-shelled rock shrimp that are trawled offshore from the east coast of Florida at a depth of eighty feet taste like bites of lobster. Royal reds, which live on the bottom at the edge of the continental shelf in depths of twelve hundred feet, are so sweet and tender that they taste more like scallops than shrimp. Brown shrimp feed on iodine-rich algae that gives them a hearty iodine flavor, while brownies from the West Coast of Mexico have a milder flavor. When pink shrimp from the Tortugas are cooked, their shells turn a deeper shade of pink than other species, and their meat turns from translucent to pink skin tones. Their flavor is excellent—almost as good as the succulent white shrimp from the Carolinas and Georgia, which, except for the northern spot prawn, are the sweetest of all. But it's all subjective and a matter of taste. If you eat them breaded and covered with cocktail sauce, you'll never notice the difference.

3

Glory Days

Because shrimp are such an abundant resource, archaeologists assume that ancient people ate them. Barbara Voorhies at the University of California speculated that the elevated pavements of small clamshells that she found in the coastal wetlands of Chiapas, Mexico were shrimp-drying platforms. They appeared much different from the regular kitchen middens, or garbage dumps where people haphazardly tossed their clam and oyster shells. She also discovered the remains of clay hearths, which she suggested were used to preserve shrimp when the sun wasn't shining.

By sifting through the dirt and shells, she uncovered the nearly fossilized vertebrae of tiny fish that are typical of shrimp bycatch in small meshed seines still used in the same area today. The catch is dried on wooden platforms when the sun is out or dried under cover when it's not. In the actual kitchen middens, they found plenty of remains of conches, clams, fish bones, turtle shells, and bones of dolphins, but no shrimp. The thin chitinous shells quickly rot away.

The evidence of ancient shrimp eating remained circumstantial until Irv Quitmyer, a zoo archaeologist at the University of Florida, discovered microscopic shrimp mouthparts in middens in Georgia dating back to 600 A.D. It was the first hard evidence that ancient people ate shrimp, most likely the white or brown shrimp that still abound in the salt marshes there today. Later, he was part of a team that found the remains of both small marine shrimp and the big freshwater prawn, *Macrobrachium*, in vases of Mayan tombs in Copan, Honduras. One tomb had a beautiful medallion of a stylized shrimp etched in gold, with long flowing antennae and delicate legs and tail fan. And across the world in Italy, archaeologists unearthed clay vessels from the ruins of Pompeii that were also decorated with shrimp.

In North America, Native Americans wove willow branches into weirs or fish traps, twisted Spanish moss or beat the fiber out of palmetto fronds to make string, wove them into nets, and feasted on shrimp, oysters, and other delights. Meanwhile, early European settlers in the New World subsisted on a boring diet of salt beef and gruel made from grain. As they waited for the sailing ships to arrive with supplies, they got sick from malnutrition and almost starved to death on one of the most protein-rich coasts in the world.

In the 1840s, a journalist described fishermen seining for shrimp in Galveston, Texas, but the market for them was tiny until the California Gold Rush, when thousands of Chinese immigrants came to the San Francisco Bay area. Many came from the delta of the Pearl River, where they had harvested small shrimp for centuries. Those immigrants turned to what they knew best—catching shrimp, *Crangon franciscorum*. About an inch long, they can be present by the billions, but they are hard to see, because they dive into the sand and vanish. The catch, mostly dried, was exported back to China or sold to Chinese communities in the United States.

By 1875 there were thirty shrimping villages around the Bay Area and hundreds of Chinese fishermen in San Pablo Bay, many using traditional junks. Between 1889 and 1892 the average annual Bay Area shrimp catch was 5,400,000 pounds, but the total catch or landings eventually declined from both overfishing and stream pollution by hydraulic gold mining in the Sierra Nevada. This caused a constant flow of mud into the bay, damaging the marsh grass where the shrimp matured.

Author Jack London described the San Francisco shrimp fishery in his book *Tales of the Fish Patrol*:

> *"...where the tide ebbs and flows, the Chinese sink great bag-nets to the bottom, with gaping mouths, into which the shrimp crawls and from which it is transferred to the boiling-pot. This in itself would not be bad, were it not for the small mesh of the nets, so small that the tiniest fishes, little new-hatched things not a quarter of an inch long, cannot pass through. The beautiful beaches of Points Pedro and Pablo, where are the shrimp-catchers' villages, are made fearful by the stench from myriads of decaying fish, and against this wasteful destruction it has ever been the duty of the fish patrol to act."*

Racial prejudice by jealous white settlers also reigned. In the early 20th century, laws, limits, and seasons were established that made it impossible to fish. All that's left of China Camp village today is a sixteen-hundred-acre state park on San Francisco Bay, a mom-and-pop restaurant run by descendants of a fishing family that sells shrimp cocktails, some reconstructed shacks, and interpretive signs on the history of the fishery.

In England and France, another closely related species of shrimp, *Crangon crangon*, is also prized. Long before commercial quantities of penaeid shrimp were discovered in the New World, the French and British were eating the little gray caridean whole in numerous dishes.

Guy de la Valdene grew up in France on the shores of the English Channel, where the sport of catching shrimp was as prized as the act of eating them. "We would go out into the surf and drag a box like dredge on a pole, then pick out the frantic little jumping crustaceans and take them home to cook. We'd bring the water to a boil, throw them in, and then put a white-hot poker into the water. It exploded into steam, and the flash-cooked shrimp were done. The air filled with the pungent odor, the cooking water was instantly dumped out, and the little crustaceans, now turned to a bright shade of orange, were ready to eat—shell, head, and all. With good French bread and butter, nothing is better!"

Thumbnail-sized caridean grass shrimp are eaten in many parts of the world. Kuala Selangor is a small fishing village on the muddy waters of the Malacca Straits of Malaysia, where the air is pungent with drying shrimp, fish, and salted jellyfish.

I (Jack) left the dock of Mr. Peh's jellyfish processing plant and traveled across the wide muddy river in a skiff to see bare-backed fishermen hauling in tons of shrimp that were less than an inch long. The fifty-foot-long funnel nets stretched out in the falling tide, straining millions of tiny grass shrimp through the tiny meshes until the nets were gorged with them. When there was barely an inch of freeboard, and their boats were about to sink, they puttered back to the dock and unloaded the catch. Women and small children spread out the shrimp and then salted and winnowed them to speed up the drying process. After they dried in the sun, turning a cheery orange, the shrimp were packed in straw baskets lined with newspapers. They were shipped to another plant to be turned into fist-sized blocks of fermented shrimp paste that are much esteemed in spicy rice dishes from Africa to India to China.

By shipping millions of pounds of dried shrimp far and wide, the Chinese in California laid the foundation for the American shrimping industry. About the time the California fishery faded out, modern industrial shrimping on the Gulf Coast and South Atlantic developed, based on penaeid white shrimp. They were so thick that linear piles or windrows of their molts were cast onto the beach at the high tide line—something unheard of today.

Cajun fishermen in coastal Louisiana had been catching and sun-drying white shrimp since the first beach seine was imported from France in 1735. In the 19th century, sun-dried shrimp were largely replaced by canneries. In the 20th century, canneries were replaced with freezers, but even today a few shrimp are still dried and sold in bars as a local delicacy, and a few old drying racks can still be found in the marshes.

Early shrimping in the Gulf of Mexico centered around Biloxi and Pascagoula, Mississippi. Biloxi schooners sailed up and down the barrier islands, tonging oysters off shallow bars and looking for shrimp. They towed smaller shallow-draft boats that were piled high with huge beach seines. When they saw white shrimp jumping along the shore and muddying the waters, crews of a dozen men jumped overboard, circled the shrimp with seines, dragged them up on shore by the ton, and sailed back into ports in Biloxi and New Orleans to unload their catch. This inspired a song:

Shrimp boats is a-comin',
Their sails are in sight,
Shrimp boats is a-comin',
There's dancin' tonight!

Tons of small white shrimp were dried and salted for the oriental market. As the demand grew, canneries began operating, but even with the sun-dried market and the canneries going full steam, shrimping remained a local fishery. Bag seines and set nets work only in shallow water, and when the season was over and the shrimp left the beaches, the nets came up empty, so the supply was always erratic, inconsistent, and limited.

For shrimp to develop into one of the world's most popular foods, it took the simultaneous development of the otter trawl, petroleum, and the internal combustion engine. On the southeastern Atlantic Coast, Sicilians laid the groundwork of the modern shrimping industry. Matteo Cannerella, a Sicilian fisherman, came to Savannah in the 1800s. Matt and his two brothers settled in the tiny community of Thunderbolt, started a small fish market, and built fishing boats similar to the ones they used in the Mediterranean. They used traditional small-meshed haul seines, traps, and cast nets to catch shrimp. Until nylon and synthetics came along, their nets were made out of linen, which had to be washed after each use and air-dried in the shade on reels so that the salt water and sun would not rot them. Most of the fishing effort was for fish, not shrimp.

Shrimp remained a local seafood until Paul Cannerella modified the otter trawl and kicked off the modern industry. The otter trawl originally came from the Netherlands and had been around for centuries, but it was built to drag on flat sandy bottoms, not the gullies, sloughs, and mud bottoms of the Georgia coast. The rigid frame snagged the seafloor and dredged up mud, not shrimp, so Cannerella removed the iron bar that kept the otter doors apart and replaced it with chain that flexed and moved over the bottom.

The otter trawl and the internal combustion engine were developed about the same time. Without the trawl, there was no practical way to harvest shrimp in volume. Without a strong-enough engine, there was no way to pull the trawl. The little gasoline engines on the early boats were not designed to pull heavy trawls, and they burned out under the strain. As the demand for shrimp increased, diesel engines were adapted for use in boats. With them, shrimp boats could travel up and down the coast, explore new territory, and drag in deeper water, pulling the heavy funnel-shaped trawl net over the seafloor like a mule pulling a plow.

Power winches were hooked to the new engines in the 1920s. This made it possible to lift the nets on board more easily, empty the nets, and keep on dragging, enabling the boats to fish with smaller crews. Fishermen learned the bottoms and developed an array of nets to fish them. Soon everyone was dragging the Cannerellas' shrimp nets and catching shrimp by the ton, but the best fishermen could get for big white jumbo shrimp was three dollars a bushel, and many went bad before being sold. Today the same bushel would sell for over $200. "Shrimp were like hotdogs. We had to build the market with publicity. It took time," said Frank Cannerella.

As vessels became larger, faster, and stronger, it became possible to explore new fishing grounds, catch shrimp offshore, and catch them nearly year-round. Smaller boats dragged the shallow bays and estuaries, while larger boats, fifty feet and more, worked offshore. Ever bigger winches, rigging, and engines put greater strain on old wooden hulls, so by the 1960s steel hulls and fiberglass boats were built that could drag heavier nets over the bottom.

The fishermen came from several different backgrounds. In the pre-Civil War South, African slaves who arrived knowing how to grow rice and seine shrimp fed themselves from the vast marshes of South Carolina and Georgia for generations. After the Civil War, their children became the fishermen of the Mosquito Fleet. The children and grandchildren of that generation became commercial shrimpers in the 20th century. Scots-Irish farmers' sons from Appalachia and Georgia, French-speaking Cajuns from Louisiana—they all fished the Atlantic and Gulf Coasts for generations. After the

Vietnam War, Vietnamese fishermen arrived on the Gulf Coast. Whole families lived year-round aboard their trawlers in Mississippi, where they were not welcomed by local Anglo shrimpers.

Most boats were family-owned. Mostly the men went to sea, and the wives stayed home and worked in seafood processing plants, but a few women ran boats. Some had all-women crews. Sometimes whole families took to sea, with husbands, wives, and the kids running large trawlers. There were also fleets of corporate-owned steel-hulled vessels equipped with freezers that enabled them to stay out roaming the seas for months until their hulls were crammed with shrimp.

Once caught, the shrimp had to be marketed. Aside from canned shrimp, it took years to convince people that fresh shrimp were fit to eat. Americans considered seafood to be mysterious and intimidating. Although brined shrimp had been sent in barrels from Fernandina Beach, Florida to New York on the first rail lines at the turn of the century, the bulk of the nation ate meat, corn, and potatoes, not strange-looking insects from the sea. Canned shrimp and oysters were considered safe, but the fresh or frozen product was unknown and considered dangerous. Shrimp, like tomatoes, were sometimes even considered poisonous.

Before 1945 most shrimp were sold fresh in coastal markets, and the only shrimp that could be sold were big white shrimp that were caught in the daytime. White shrimp have a white translucent color and vivid green tail fins. Pink and brown shrimp are somewhat darker, and that caused buyers to assume that they were spoiled. At dark, fishermen had to quit trawling, or their nets would fill with the unmarketable brown and pink shrimp.

After World War II everything changed. People became more adventurous. They had seen the world, and they weren't content to stay in one place. Families drove to Miami Beach and New Orleans and discovered the delights of seafood. Eating shrimp was a bold and adventurous thing to do. When they got home, they bought canned shrimp in the grocery stores.

When Clarence Birdseye perfected a process for selling frozen foods, shrimp took off. Once frozen and breaded, it didn't matter what color the shrimp had been when they were caught. Brown and pink shrimp became as valuable as whites. In the late 1940s, large concentrations of brown and pink shrimp were discovered in the Gulf of Mexico. Shrimping spread up the Atlantic Coast to the Carolinas and as far north as the southern end of Chesapeake Bay. With refrigeration and better highways, inland cities enjoyed seafood restaurants, where you could get oysters and clams on the half shell, broiled lobsters, and a plate of boiled shrimp.

As frozen shrimp began to sell, and people found out how delicious shrimp were, the demand grew, and so did the size of the vessels. Originally shrimp boats dragged just one net. In the mid-1950s, double-rig trawling was tried, in which a pair of nets were pulled off long booms called outriggers on each side of the vessel. This technique was quickly adopted by most larger offshore boats.

Pink Gold was discovered in 1949 when a shrimp boat strayed into the waters west of Key West and bogged down with unbelievable quantities of large pale pink shrimp. It had been suspected that a mother lode was out there, since grouper, snapper, and shark stomachs were often filled with shrimp. Hundreds of tons were harvested from the white marl bottoms, soon to be known as the Tortugas fishing grounds.

Government fisheries scientists worked hand in hand with fishermen to develop the shrimp fishery. In the 1950s the National Marine Fisheries Service engineers experimented with electric tickler chains that shocked the shrimp out of the mud and into the net. They discovered new shrimping grounds such as the big juicy royal red shrimp off Alabama in two hundred fathoms. Some boats ventured out to the deep water at the edge of the continental shelf to harvest royal reds in one thousand feet of water.

By the late 1960s, government food technologists developed new ways to preserve shrimp at sea and keep them longer by using sodium metabisulfite dips, which are now considered a health hazard. Federal and state biologists conducted tag-and-recapture programs to study the movements, migrations, and growth of shrimp. And they tried to manage the fishery by closing bays and nursery grounds until the shrimp grew larger and achieved legal size. Shrimp were discovered off South America, and American shrimpers flocked to the Guianas in the 1960s. Corporations invested millions in fleets of boats. Processing plants sprang up in Honduras and Nicaragua, shipping shrimp back to Miami by the ton on freezer boats. By the mid-1970s, catches declined and American investors bailed out, but the Japanese government moved in to start a cooperative venture with the Suriname government.

In the late 1970s, we were in Suriname on a collecting expedition for freshwater fish sponsored by a trustee of the New York Aquarium. The Minister of Fisheries of the Republic of Suriname invited us to go along on an inspection cruise aboard a new steel-hulled Japanese trawler. As we headed down the Marowine River, we passed abandoned American shrimp boats. Some trawlers were sunk at the dock, and others had been there so long that trees and bushes were growing on their decks.

Although the freshwater rivers of Suriname abound with strange and unique creatures such as armored catfish, electric eels, and piranhas, life is pretty much the same from one big muddy estuary to the next. So when the nets were hauled back, the catch of shrimp, crabs, and finfish looked like Louisiana. Our Japanese hosts turned them into a magnificent sashimi served with lime juice and sauces.

In Port Canaveral and Apalachicola, Florida, fishermen quit culling the delicious rock shrimp overboard and started selling them in the 1970s. Pacific Coast shrimp boats strayed north as far as Canada and Alaska and discovered a bonanza of delicious northern pinks. These were several species of the genus *Pandalus*, caridean shrimp that carry their eggs on their abdomens, instead of broadcasting them into the sea as penaeid shrimp do. Across the continent in Maine, another species of cold-water shrimp, *Pandalus borealis*, was discovered, starting a small commercial fishery, but the mainstay of the industry always remained pink, brown, and white shrimp in the southern shallows.

On the U.S. Atlantic Coast, Fernandina Beach near Jacksonville became a major shrimp port, as did Savannah, Charleston, and Beaufort, North Carolina. Ninety-foot-long steel-hulled trawlers and even larger vessels pulled four and even six nets at a time. Those nets evolved: flat nets, balloon nets, skimmer trawls, butterfly and mongoose nets—each designed to work in different environments and catch different species, and all trying to outsmart the minuscule brain of a shrimp.

It takes a lot of people and organized efforts to bring shrimp to the table. Whole towns and economies were built on these little crustaceans. When the shrimp ran, cash flowed. But when there was a bad year, poverty reigned. To know when the last big shrimp run occurred years ago, you could just look at the model year of the shrimpers' pickup trucks and the Cadillacs and Lincoln Continentals of the fish-house owners.

When the boats dock at the processing plants, shrimp are shoveled out of the holds onto conveyor belts, and nearly anybody who's willing to work in the pungent, forever wet, and often cold environment can get a low-paying job. The old-style fish houses where shrimp were unloaded, iced, packed into boxes, and loaded onto trucks were replaced with "value-added" plants where they are peeled, deveined, breaded, and even cooked.

Nothing in the world is counted more carefully than shrimp, except money. They are sold by the number of shrimp per pound. The larger the shrimp, the lower the count and the higher the price. Producers are careful not to mix small shrimp with big ones, because doing so "spoils the count." The difference between 21\25s (twenty-one to twenty-five shrimp per pound) and 26\30s (twenty-six to thirty shrimp per pound) can mean that a boat will make or lose thousands of dollars on a trip. So shrimp are counted over and over with total intensity, on the boat, at the processing plants, on farms, in seafood markets, and at restaurants. Names like "jumbo, medium, and small" are retail marketing terms that have nothing to do with how shrimp are actually bought and sold. Restaurants serve dishes of tiny peeled and breaded "popcorn shrimp" that count eighty or more to the pound. Truly monster shrimp that run less than ten to the pound, called "colossal," are rarely seen in fish markets because they are whisked off the boats by Asian buyers. To the Chinese and Japanese, bigger is better, be it a melon, jellyfish, or the grandfather and grandmother shrimp.

An uneasy relationship exists between the fishermen who catch the shrimp and the processor who sells them. He who owns the fish house controls the price. When the catch is weighed and sold, he who catches the shrimp believes that he's been peeled, deveined, and screwed. But since the shrimper and the processor need each other, they must coexist in a symbiotic relationship.

As the industry grew, channels in southeastern ports were dredged and deepened to accommodate deep-drafted trawlers as the boats became larger. High-rise bridges were built across inlets and bays so that shrimp boats' masts and raised outriggers had clearance to pass beneath. In the 1970s the government gave low-interest loans to commercial fishermen so that they could buy larger, modern trawlers built of steel and fiberglass with bigger engines. This enabled them to go farther, weather the seas, and stay out longer. Every year, better electronics, sonar, radar, and GPS made the fleets more efficient. Nobody gave any thought to overfishing or what impact all those nets scraping the bottoms might be having on other marine life.

By 2003, shrimp replaced canned tuna as America's best-selling seafood. "There's a million ways to cook shrimp," Bubba Blue told Forrest Gump in the movie of the same name. "Shrimp is the fruit of the sea. You can barbecue it, boil it, broil it, bake it, sauté it. Dey's, uh, shrimp-kabob, shrimp creole, shrimp gumbo...." Every culture or geographic area has a shrimp recipe of its own, with an endless array of spices, boils, dips, and sauces (see Figure 3.1).

Figure 3.1 Peel-and-eat shrimp plate

Courtesy of the Florida Department of Agriculture

The secret is to toss fresh whole shrimp into a pan with a half inch of beer or water, just enough to keep the shells from burning, and add spices to taste. Give it a slow heat for three to five minutes. Their heads have enough water to cook them and produce a delicious stock. Beware of overcooking, especially with royal reds.

While deep-water royal reds and Pacific pinks look cooked while still alive, shallow-water pinks, browns, and whites turn red only when they are cooked. Watching fresh shrimp in boiling water rapidly change color from brown or white to red before your eyes is very dramatic. The magic is caused by a protein called astaxanthin, a much more powerful antioxidant than vitamin A, which gives cooked shrimp, crabs, and lobsters that appetizing rosy pinkness. Astaxanthin is the same molecule found in plankton, krill, and algae that gives northern lobsters and crabs their blue color. Exactly how the color changed remained a mystery until chemists described electrons from astaxanthin bonding with other protein molecules during the cooking process and showed how that affected the absorption of light. Both crustaceans and salmon get their color from eating plankton, krill, or algae.

Shrimp 'n' grits is a South Carolina favorite, as is jambalaya in Louisiana. Asia has endless shrimp and rice dishes, including oat shrimp, served in Thailand. The Italians serve them with spaghetti, and the Portuguese make shrimp tempura. Add to that soft-fried shrimp, batter-fried shrimp, sake steamed shrimp, drunken prawns, steamed fresh prawns, and the list goes on and on. There are indeed a million ways to cook shrimp and another million ways to season them.

Shrimp are one perpetual party, and endless shrimp and seafood festivals celebrate them (see Figure 3.2). There's a rock shrimp festival at St. Mary's, Georgia, and a Wild-Caught Shrimp and Grits Festival on Jekyll Island. Louisiana has its Shrimp and Petroleum Festival. Florida has its Eight Flags Shrimp Festival on Amelia Island, the Everglades Festival, and the Seafood Festival in Apalachicola. From the Carolinas to Texas, seafood festivals happen in nearly every coastal town, but Alabama's National Shrimp Festival arguably is the biggest. Smells of frying shrimp, soft-shell crab, fish, and hush puppies mingle with cotton candy and funnel cakes. Guitars and banjos twang, harmonicas play, entertainers sing country-and-western songs, and children listen to storytellers. People heap plates with food, mill about, pick over trinkets at the arts-and-crafts stands, and watch skydiving, balloons, and fireworks.

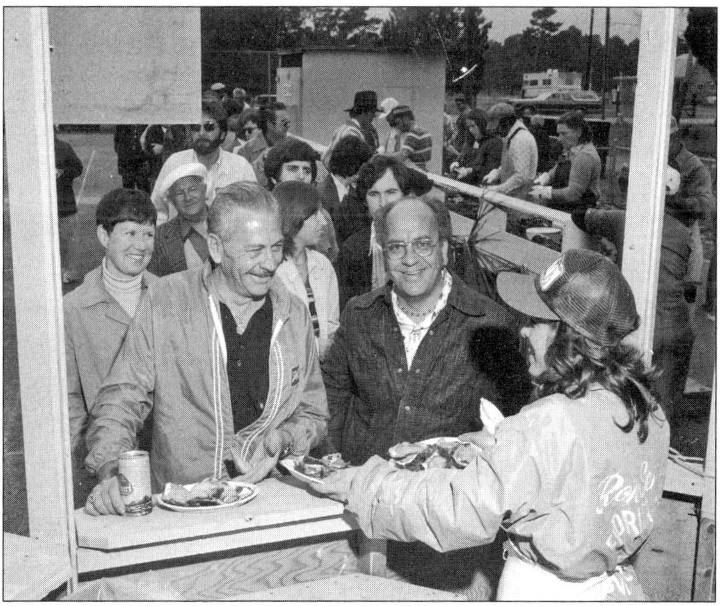

Figure 3.2 A Florida seafood festival
Courtesy of the State Archives of Florida

A Blessing of the Fleet often occurs (see Figure 3.3). Because shrimping is so precarious, and everyone in the community knows someone who was lost at sea, wreaths are cast into the water in their memory amidst prayers for safety. For visitors it is a colorful tourist attraction, but for fishermen, it is deeply meaningful.

Figure 3.3 Blessing of the Fleet in Key West

Courtesy of the State Archives of Florida

An economist conducted a survey for the *New York Times*, trying to fathom why American consumption of shrimp jumped from 1.4 pounds per person in 1980 to 4.1 pounds in 2007. Was it due to the rise of ethnic foods, the opening of many more Chinese and Thai restaurants, a change in demographics, decreasing prices, or even the popularity of the movie *Forrest Gump*? It turned out that a good part of the demand that sent trawlers scouring the sea worldwide and caused coastlines to be chopped into shrimp farms was created by seafood restaurant chains. Restaurants looking for fast, convenient, and tasty dishes promoted shrimp. Red Lobster, Captain D's, and others have saturated the nation with television commercials depicting happy people eating shrimp, until America has become a shrimp-eating nation.

As chain restaurants proliferated, fast food got faster. People quit cooking at home as fewer women stayed home. People began eating more shrimp than ever because it fits into busy lifestyles. A working woman who has to fix dinner for the kids grabs a bag of shrimp from the freezer, tosses it into a pot, and in minutes has a tasty and nutritious meal. It is the ultimate convenience food. No longer a specialty food, shrimp were available to the masses, including those who didn't live near the coasts. Whole cookbooks were devoted to shrimp, and dishes proliferated in seafood restaurants. Americans eat an astonishing 963 million pounds each year, but no one subsists on shrimp. Eating shrimp has more to do with lifestyle, ambience, and state of mind than protein.

Years ago, when people wanted to splurge, they went to the grocery store and bought an expensive packaged shrimp cocktail. Now, thanks to shrimp farming, shrimp are nearly as cheap as hamburger, but people still think of them as a luxury food. Mountains of rosy pink jumbos are heaped upon buffet tables at opulent casinos in Las Vegas and artfully displayed on platters at wedding receptions. Now so many types of shrimp are available from oceans around the world that, in addition to the usual pink, brown, and white shrimp, a hostess can dazzle guests by serving deep-water royal red shrimp, prawns from the Pacific Northwest, or rock shrimp.

Sam Dunlap, who runs the Seineyard, a popular seafood restaurant in Woodville, Florida, said, "To me shrimp is a miracle food. They don't spoil unless you're negligent, they taste good, they're easy to prepare, and they're one of the most profitable items on a restaurant menu. It doesn't take a gourmet chef to cook them; you just thaw them out and boil, steam, or bake them. We have fifteen to seventeen cents in a shrimp. We add a few garnishes to the plate, and our cost is two to three dollars, but we can charge eleven. You can't do that with other seafood. You need a 300 percent markup to make any profit in the restaurant business, and it's not there with most fish, but it is with shrimp."

Not only are shrimp good to eat, but they are good for you. Shrimp are a rich source of protein, vitamins, and minerals and are low in fat. Cholesterol levels are lower than in chicken and beef, making it ideal for weight loss or weight maintenance diets. They have never been implicated in human diseases. Wild-caught shrimp are a good source of omega 3 fatty acid, which medical research shows may reduce the risk of heart disease.

Shrimp feed at the bottom of the food chain, eating diatoms, detritus, and bottom-dwelling invertebrates. While large predatory fish such as sharks, swordfish, and tuna are long-lived and accumulate heavy metals in their tissues, short-lived shrimp normally do not accumulate heavy metals like mercury unless they're held in pens in polluted bays. Furthermore, because they grow so explosively and reproduce so fast, there's little danger of their being overfished, unlike cod, haddock, and many other fish.

Despite almost a century of intensive commercial trawling, inshore shrimp have not been overfished to the point of depletion. Studies by National Marine Fisheries Service scientists showed that an otter trawl passing through shallow-water shrimp takes only one in sixteen of the shrimp present. Although total landings have declined somewhat in recent years, it is the economic pressure of high energy costs and cheap imported shrimp, hurricanes, and stricter environmental regulations that have driven fishermen off the water.

As the demand exceeded the supply of wild-caught shrimp, and environmental problems of drowning sea turtles and bycatch got more attention, shrimp farming developed. Farmed shrimp are now cheaper than crab and a great number of fish. They can be grown in great quantities and are not subject to the intensive regulations that have closed or reduced many wild fisheries on the U.S. East and West Coasts.

4

Shrimp in the Grass

As I (Anne) paddled a kayak out from shore into the sea grass meadows of St. Joe Bay in the Florida Panhandle, it was a warm summer morning. A bald eagle was a pinpoint in the blue forever overhead. As always with eagles, it was so high that it took me a while to be sure it really was an eagle and not an osprey. A marine biologist, I had studied and taught university students about the sea grass ecosystem for many years, but today I was in search of a mysid shrimp that was so small it was nearly invisible as a baby and hardly any bigger as an adult.

The goal was to begin a scientific shrimp farm. The species was used to test the toxicity of oil drilling effluent. In order to get a permit to dump their waste from offshore drilling platforms, drillers had to provide the government with data proving it didn't exceed specified toxicity levels. This was obtained by exposing test shrimp to the effluent and then monitoring their survival, growth, and reproduction to make sure everything remained normal. The best place to get them was in the sea grass meadows. Wild ones were not used for actual testing, but they could be used to build a breeding colony to provide the parent stock for the newborn shrimp that the test regulations required.

Sea grasses grow, flower, and seed in clear shallow water, with currents rather than wind or insects to scatter their seed. Along the Atlantic and Gulf Coasts of the United States, they occur inside bays from New England through the Chesapeake Bay to Florida and around the Gulf of Mexico to south Texas. One of the largest meadows in the world occurs on the Gulf Coast in Florida Bay, offshore from the Everglades and from Clearwater north to St. Marks, Florida, where grass grows from the intertidal zone to fifty miles offshore. Sea grasses are not so widespread on the Carolina Coast, or in the vast muddy estuaries of Louisiana, Georgia, and Texas. The murky water makes the seafloor too dark for rooted plants to survive. Although sea grass thrives inside bays from New England through the Chesapeake Bay, it's too cold for commercial pink shrimp to grow in them. North Carolina, where the water is clearer and shoal grass finds enough light to grow, has enough pink shrimp to support a small fishery. Sea grasses are also widespread along the Pacific Coast of the United States.

Perhaps no ecosystem on Earth is so easy to reach, so full of animals, and yet so little known to humans as a sea grass meadow. Larger species, such as fish and sea turtles, quickly move out of a diver's visual range when a noisy flailing human splashes down into their world. Smaller animals remain motionless, camouflaged and invisible.

Southern sea grasses are full of both edible pink shrimp and tiny species of mysid and caridean shrimp known only to marine biologists. The large pink shrimp are nocturnal and remain completely buried in the sand during the day. The grass beneath the kayak teemed with caridean shrimp of many different species, but I couldn't see them because they were tiny, transparent, or embedded in the dense coating of algae growing on the blades of grass. Although most carideans are minuscule, less than an inch long, they have more species and more ecological diversity than nearly any other group of shrimp. They are found all over the world at all latitudes and all depths, on soft or hard bottom. Different species may be free-swimming, freshwater as well as marine, and are mostly nocturnal.

The grasses seethe with life, but only a few creatures, like sea urchins, manatees, and green turtles, actually eat the unpalatable grass itself. Most creatures, like the pink shrimp, graze instead on hundreds of species of delicate hair-like or encrusting algae, diatoms, and animals that attach themselves to the blades. By their tenacity and sheer numbers, these little shrimp become the largest dominant decapod predator in turtle grass. Their huge populations scour the roots, stems, and leaves, devouring algae, mollusks, crustaceans, and fish fry. If they and other grazers didn't mow down the algae, it would hurt the ecosystem by blocking sunlight and choking out the grass.

The shrimp and snails that eat tiny plants must be tiny as well, and that makes them vulnerable to the hordes of hungry fish that are everywhere. The evolutionary pressure is to stay hidden, and it is not immediately obvious what is going on. What appears is totally a function of the net's mesh size. The shrimp that swarm through the meadows by the millions are tiny, nocturnal, or both, and thus are literally invisible to a human snorkeling about during the day.

A bizarre green needle-shaped caridean shrimp called *Tozeuma* has a long pointed head and is so camouflaged that there can be thousands in a square meter of grass, but you'll never see one until you swoop through with a fine meshed net. It stays on the grass, its needle-thin body aligned on the blades along with tiny baby sea horses and pipefish that lurk in the jungle, ambushing the shrimp where grass bed meets open sand.

Needle shrimp can swim sideways, looking like a broken piece of drifting shoal grass. They rarely get eaten by the larger pink shrimp. But not so for the many tiny species of caridean shrimp that have scientific names like *Thor*, *Hippolyte*, and *Periclimetes*. One is bright green with lemon yellow eyes. Another species can change colors from green to brown to match the surface it is on. Even though they can change their body colors from brown, golden, or green to match the bottom, or are so clear they're practically invisible, they are among the major prey items of pink shrimp.

Anchoring the kayak, I dove into 4 feet of clear calm water over the grassy bottom and snorkeled off. Green grass dissolved into green water in the near distance. Visibility was perhaps 15 feet. The morning sunlight swept over the meadow in waves, a mesh of golden fire created by surface ripples and wind a few feet above. Scallops sat scattered about, their brown and white shells wide open to reveal shining blue eyes. Red starfish with white polka dots wrapped their arms around the bases of the sea grass plants like children hugging their mothers. Tiny blue-eyed hermit crabs scrambled over the swaying green blades of turtle grass like squirrels in trees, hanging on as passing waves rocked them back and forth. Inside empty seashells, dwarf octopus hid, waiting for darkness to hunt the hermits. Suddenly a loud "pop" reminded me that not all small shrimp are delicate and helpless. Under a clump of drift algae, a two-inch-long greenish-black pistol shrimp lurked (see Figure 4.1). It looked like a miniature Maine lobster with one oversized claw. It could cock that claw and blast out a high-pressure jet of water strong enough to drive away both predatory fish and rival pistol shrimp. Pistol shrimp have been known to shatter the glass in a small aquarium. High-speed photography has shown that they create microscopic bubbles that collapse in on themselves, explode, and generate a flash of light and heat approaching the temperature of the sun.

Figure 4.1 Pistol shrimp

Courtesy of Gulf Specimen Marine Laboratories, Inc.

At the edge of the white sandbars that lay scattered throughout the grass bed, endless schools of transparent mysid shrimp, each no longer than a baby's fingernail, hovered. Also known as "opossum shrimp" because they brood their babies in pouches, mysids are often used as aquarium fish food, because even the most expensive finicky eaters, like sea horses, angel fish, and even sea dragons can't resist them. There must have been literally billions in the bay. Each shrimp faced into the current, darting back and forth to seize even smaller creatures that swept past, crustaceans too small for the human eye to even see. The shrimp intermingled freely with huge schools of equally transparent, equally small baby fish.

I went back to my anchored kayak and got a net designed to collect the smallest things visible to the eye. I towed the small beam trawl with a very fine mesh net by wading through waist-deep water with the tow rope wrapped around my hips, plodding along like a mule pulling a plow. Catching uncountable numbers in a few minutes, I poured them into a tray of water for a closer look. Most of the mysid shrimp turned out to be one species. Miraculously, it was the species I needed. Mixed in were a few other species of mysids. Some had huge black eyes and looked like tiny raccoons. Others had elongated needle-shaped bodies. What did these rarer mysids do differently from their explosively abundant cousins?

I picked out the shrimp that weren't the right species and let them go. The rest went into a chest of seawater to take back to the lab. I climbed back into the kayak and headed in as a powerboat sped past. It could get to farther-out places, places I couldn't reach that afternoon with my slow pace and the brief time I had. But the people in the powerboat missed the nuances—how the wind was gusting and dying and gusting, the blazing sun in each drop of water as it blew off the paddle blades. Now two eagles were soaring overhead. They banked and turned, soaring higher and higher, the white heads and tails flashing and disappearing and flashing again. After a few minutes the sky absorbed them, and they were gone.

In the following months, the mysids in my tiny farm flourished. Their populations grew explosively until it looked like our Gulf Specimen Marine Lab might be able to become the main supplier in the country. Week after week I worked—designing and building a system to rear them on a large scale, developing innovative ways to harvest the tiny newborns from the huge tanks, figuring out when and where to find them in the field.

Every time the project seemed to be going smoothly, a new challenge appeared. As soon as I solved one problem, another developed. The tiny shrimp were always mixed together with equally tiny baby fish. Learning how to efficiently separate fish from shrimp was a challenge that I never quite mastered, and the fish turned out to be unbelievable shrimp eaters as they got a little bigger than the mysids. But with each new issue, I learned and learned, and not just how to culture this particular species.

The main collecting site was a two-hour drive away. I needed someplace more convenient and started checking up and down the coast wherever there was road access to the water. A five-minute pull of the net at a place just a few feet from a highway yielded literally hundreds of thousands of mysids and innumerable other small creatures. Only a few days before, hardly any were there. As the trucks roared by, I kneeled in the sand, staring into a bucket whose water was black with the tiny creatures. Large half-grown pink shrimp were everywhere as well. It would be a good fishing season in a few more months.

A month later, the population in one of the six tanks collapsed. All the shrimp died overnight. The others were at about the same density as before. As always, just as the project had seemed to be on the verge of stabilizing and becoming a source of income, something else had happened. Usually it was due to something I could recognize and fix, but this time there was no obvious reason. It seemed to reflect the boom-and-bust pattern in the ocean—the mysids would appear explosively for a while and then be gone again. Nobody knew why. But each time it happened in our tanks, it would take weeks and months before the population would be large enough to support a harvest again. In the meantime, feeding them with brine shrimp was a twice-a-day, seven-days-a-week commitment. If they missed a feeding, they'd eat each other.

Like shrimp farmers everywhere, I realized that it was one thing to study life in the sea, but it was another matter altogether to try and manipulate it to meet some human goal. All the ups and downs of farming, both physical and emotional, were becoming exhausting. I tried not to take each setback seriously, and I vowed to keep trying. I fixed the filter once more, or whatever it was this time. And each sale gave me new energy to keep going.

By the end of the summer, our Gulf Specimen Marine Laboratory finally had sold enough mysids to pay back the initial start-up costs, and we broke even. But fall weather came suddenly, the temperature dropped, and we lost half the animals in one night. To install heating would be expensive, and there would be no further earnings for months until the colony rebuilt. More months of unpaid seven-days-a-week work.

In the end, the project failed. The adults, used to the semi-natural environment I had developed in my system, didn't reproduce well in the artificial environments they faced after being shipped to other labs. Then the regulatory requirement to use this species for effluent testing was dropped, and the market vanished overnight. It was finished.

The mysids had presented some tough lessons and raised Big Questions, the fundamental issues of life and death as the populations boomed and disappeared repeatedly in the culture tanks. What did all this endless recycling of generation after generation of shrimp, the turning of the life force in the culture tanks, really mean? How did the ocean explode into shrimp where there had been only one or two? I hadn't really answered any of those questions, but they had come into focus in a way they never had before.

Whenever I towed the mysid net at night, I always caught a lot of half-grown pink shrimp. As the last rays of sunlight fade away, pink shrimp emerge from their blanket of sand to eat both the algae and the smaller caridean shrimp embedded in the algae. When the tide is low and the grass flats along the shore are exposed, wading with a gasoline lantern reveals the ghostly little creatures, their eyes glowing like tiny hot-pink stars under the blades of turtle grass. They freeze under the glare of the lantern, and it becomes clear what beautiful creatures they are. Graceful and agile, they seem more spirit than flesh. They perch on the soft bottom with thin, delicate legs. Then, using the swimmerets beneath their tails to churn the water, they lift off the bottom and swim, using their flat tail fins as a rudder. By lifting one flap and lowering the other, they turn on a dime and flip away the instant a dip net closes in, jumping over the rim and vanishing into the darkness.

In the sea grasses at night, lit only by starlight or the moon above, pink shrimp swarm over the clumps of algae and climb up the grass blades to munch on whatever algae and small invertebrates come before them. They crush the exoskeletons of amphipods, isopods, tiny crabs, and the shells of little clams in their mandibles. Pinks pick among the thin, wispy leaves of Cuban shoal grass, seizing multilegged polychaete worms. Sausage worms and ribbon worms are fair game too, as are the tiny crabs, but the shrimp disregard the orange starfish and brittle stars and shun the toxic and bad-tasting sponges.

Penaeid shrimp eat everything, although it's difficult to say exactly what. Their stomachs, which are lined with spines called "stomach teeth," are so efficient at grinding up their food that it becomes unrecognizable. Over the past century, persevering scientists in various locations from North Carolina to Texas have teased apart the digestive tracts of shrimp, examined the contents under microscopes, and found a potpourri of bottom-dwelling small animals. Shrimp are omnivorous, eating shelled amoebas called foraminiferans, tiny snails, squid, small fishes, plant material, and debris, as well as worms. In Tampa Bay sea grass meadows, pink shrimp consume dinoflagellates, nematodes, polychaetes, ostracods, copepods, mysid shrimp, isopods, amphipods, caridean shrimp, and their eggs. Woe unto an inch-long baby flounder or a larval transparent tarpon if it encounters an eight-inch jumbo shrimp. But they never lose their taste for the tiny algae that grow on the blades of grass.

With a hydrophone, scientists have recorded the distinct sound of pink shrimp grazing on the tiny algae that grow on the surfaces of the grass blades, tearing apart their food. Amplified, the grazing shrimp sound like a bear tearing bark off a tree. The shrimp jump back and forth from the grass to moving seaweed clumps as they're dispersed by tides and currents, chowing down heavily on diatoms and filamentous algae that encrust both the seaweed and grass blades. Pink shrimp attack the many species of tiny caridean shrimp that live in the clumps of drift algae. At two to four inches in length, young pink shrimp are gigantic compared to the mysid and caridean shrimp. They dominate and thrive amid the jungles of vegetation.

Shrimp, along with tiny herbivorous snails, are the stewards of the sea grasses. Without their continuous grazing on the encrusting algae, it can become so dense that it shades out and kills the sea grass blades on which it grows. Small pinfish eat the snails and shrimp and are food for predatory game fish like sharks, gag grouper, and redfish. Intense recreational fishing for grouper and redfish allows the shrimp-eating pinfish to explode in numbers. The pinfish then eat so many of the tiny algae-grazing shrimp and snails in the sea grasses that the algae can take over, especially if the water in the area also has excess nutrient levels.

Because so many game fish love to eat shrimp, juvenile pink shrimp are harvested from sea-grass beds and sold as live bait from the Carolinas to Texas. The live-bait shrimp industry, centered largely in Florida and Texas, is worth millions and is essential to the billion-dollar-plus sports-fishing industry.

At age fifteen, when I (Jack) had moved to the little fishing village of Carrabelle, on the Gulf side of Florida, I had insatiable curiosity about everything that swam or crawled, burrowed or flew. Eager to see what I could catch in the river, I bought my first cup of frozen bait shrimp. The shrimp could have been any one of four species—whites, pinks, browns, or sea bobs.

All I had ever seen while growing up in New York City were headless shrimp in the fish market. The shrimp stood apart from the clams, oysters, and fish in the showcase. They weren't crabs or lobsters, although it was clear that they were somehow related to them. They were encased in paper-thin, almost transparent segmented shells with red flaps on their underside. I always wondered what the rest of the animal looked like, expecting it to be some sort of lobster or crawfish.

After I thawed out my overly ripe bait cup, the mystery was solved. When I saw what the whole shrimp looked like, it was so creepy I almost wished I hadn't. They resembled bizarre underwater insects. Their long head spines were serrated and sharp, their eyes bulbous and stalked. Their long, thin legs ended in tiny pinchers. I extracted them by their soggy and repulsive tangle of whiskers, and they looked like the hair of something very dead. No wonder people broke off their heads! Nevertheless, I rammed my hook into the shrimp's body and cast it into the river, and it wasn't long before I felt a snatch on my line. The white trout were running that day, and it took only an hour or two to fill a cooler with them.

When the industry began on the east coast of Florida around 1915, it was small, with a few dozen people using push nets to catch live white shrimp. A few part-time fishermen collected enough for their own use and sold some to the local country stores that sold bait and tackle. In addition to gas, beer, soft drinks, crackers, and a bit of fishing gear, you could buy crickets, earthworms, and, of course, shrimp. They kept them alive in float buckets attached to docks, but the technology of keeping them in and selling them out of tanks did not exist.

After the great land development boom crashed in the 1920s and real estate developers were sent packing, Florida's state government did everything it could to bring back the Yankees to boost the sagging economy. It advertised Florida's sandy beaches, warm climate, and sunshine in magazines and even promoted the torrential rains that flooded the swamps as "liquid sunshine." Brochures showing happy sports fishermen, holding up big speckled trout or standing beside huge tarpon, sawfish, or sharks, were distributed far and wide.

The promotion paid off. After World War II, visitors began pouring into Florida, creating a demand for bait shrimp that soon exceeded the supply. A few innovative retailers struggled with tanks and aerators, trying to keep the shrimp alive. As the demand for live bait shrimp increased exponentially, other baits fell into relative neglect.

When the market for bait shrimp exploded, several thousand people went into the bait business to meet the demand. They dragged the bays, sounds, and creeks in small boats on the east coast of Florida with small otter trawl nets for white shrimp, doing their best to keep the shrimp alive in tanks. They shoved push nets over the shallow tide flats, used dip nets, and threw cast nets off bridges and the shores of inlets. But when the fish were running, and the sportsmen lined up at the bait shops, they couldn't meet the demand.

The state dispatched a fisheries biologist to survey the entire west coast in 1949, looking for new populations of white shrimp, the main species sold in bait stores at the time. Since white shrimp are active in the daytime, he did all his sampling during the day, but his trawls always came up empty. White shrimp don't live in Southwest Florida. And they don't live in sea grasses. And he didn't know that pink shrimp were nocturnal. That difference caused him to miss an entire state's worth of pink shrimp. Had he gone out at night, and waded around the grass beds, he would have seen them all over the place.

Eventually, fishermen discovered pink shrimp by fishing at night in sea grass meadows. People dipped netted hoppers using kerosene lanterns at night to stun them, but their lights weren't very effective. The invention of the gasoline lantern changed all that. People could walk the flats, stun shrimp in a blaze of light, and scoop them up with dip nets. Catching a few dozen was easy, but a more efficient technique was needed to meet the growing demand.

The sea grasses on the west coast of Florida present an almost impenetrable barrier to catching pink shrimp, for either food or bait. The traditional otter trawls, used to catch white and brown shrimp in the muddy estuaries of the south Atlantic, simply clogged up with drift algae, sea grass, and sponges and were almost impossible to lift, much less sort through.

Roller-frame trawls were welded together by ingenious commercial fishermen who built them in their garages, testing to improve their efficiency until they rolled smoothly over the grass beds. They consisted of one or more slotted rollers attached to the bottom of a metal framed net that combed the shrimp out of the grass with a grid of stainless steel finger bars spaced an inch apart. The bars prevented the clumps of drift algae, sponges, and large fish and turtles from entering the net. As it rolled over the bottom, it let in the shrimp, crabs, and small fish and kept out the "trash."

Each night trawlers raised and lowered their trawls every ten minutes and dumped thousands of live flipping and jumping shrimp on their sorting tables, kept them alive, and sold them to fish camps and bait shops. When the trout were running, recreational fishermen lined up to get their buckets of live shrimp. Eventually the pink shrimp, *Farfantepenaeus (=Penaeus) duorarum*, became the dominant bait. They were more robust and stayed alive longer than white or brown shrimp, so people quit buying from the east coast.

Some say the roller-frame trawl was invented in St. Petersburg's Boca Ciega Bay. St. Petersburg Beach was little more than a wild sandy peninsula at the time, but it was the mecca of recreational fishing. The glassy clear waters of Boca Ciega Bay teemed with speckled trout, redfish, mangrove snapper, and other species. Because the trout fed on the huge populations of pink shrimp that abounded in the turtle grass meadows, it was known throughout the country and world as a fishing paradise.

But in the 1950s and 1960s, huge chunks of publicly owned bay bottom and coastal lands were auctioned off by the Florida state government to real estate developers who dredged, drained, and reclaimed them. Day and night for years the dredges churned, sucking up the bay bottoms to create dry land on narrow canals. The once-grass-covered shallows were soon covered with Florida dream houses, each with a seawall and a dock where sharks, sea turtles, and manatees once swam.

Bait shrimping was and still is a cottage industry of small family-owned operations that had and still have no political clout. There are no giant corporations producing and marketing live bait to lobby the government and protect their interests. The bait shrimpers and fishermen were forced to move away from the urban sprawl. Some settled in Homosassa Springs, and others set up bait shops in Crystal River, farther north on the Florida Gulf Coast. Some went to Keaton Beach in one of the most remote areas in Florida, surrounded by pine trees and swamps, in the heart of what is now known as the Big Bend Sea Grasses Aquatic Preserve.

Steve Moody ran a small fleet of boats in Keaton Beach and drove the tank truck that delivered live shrimp to the bait-and-tackle shops up and down the Florida panhandle. After they dropped off their bait shrimp at the two small shops in Panacea, they stopped at our small public aquarium to sell us octopuses, sea horses, pistol shrimp, and other specimens that came up in their nets.

Occasionally, the two of us would drive to Keaton Beach and go shrimping with Steve and his deckhand Joe. Bait shrimping was ten times harder than catching food shrimp. Keeping shrimp, or any other marine creature, alive is a lot more work than throwing your catch on ice and getting a nap or rest between tows. Bait shrimping involves changing water, eliminating waste products, and watching the temperature, making sure it doesn't get too hot or cold. From the moment you get on one of those boats, it's nonstop work from sunset to sunup. The thirty-foot diesel-powered fiberglass boat pulled two roller frames, and every fifteen minutes the two power winches raised and lowered the two roller frame trawls. The minute they were raised, their contents were shaken out onto the sorting table. The end of the net was hastily retied and dropped back for another tow.

The tows were necessarily short, because the nets quickly filled with seaweed, clogging the grates and keeping the shrimp from getting in. In the spring and fall, the amount of biomass was staggering. The shrimp were tossed into live-bait wells that lay behind and beneath the sorting tables. All the other marine life—sea horses, sea urchins, hermit crabs, fish, and sponges—was immediately thrown back into the water alive.

With blinding speed, Steve and Joe went through the catch in four to eight minutes, picking out the frantic jumping shrimp and watching out for moray eels and toadfish that could inflict painful bites. No sooner was the culling table swept off, and the shrimp put away in tanks, than it was time for the trawls to be raised again, and another sorting. The only rest we got came around midnight, when we stopped to eat and watched the lights of the other boats working out in the bay. We trawled from sunset to sunrise or as long as we were catching shrimp.

At daylight we went back to the dock with the vats crammed with shrimp. For bait shrimpers, sleep was at a premium, especially in summer, when the heat made snoozing difficult. The boats and gear were always breaking down and had to be repaired before it was time to go again, so sometimes Steve and his deckhands didn't sleep at all. The wear and tear on gear, vessels, and vehicles carrying the weight of seawater and shrimp was tremendous.

When the trout or redfish were running, the wholesaler's phone rang off the hook. As soon as his boats came into the dock, the shrimp were scooped into tank trucks that went hundreds of miles up and down the coast, delivering the shrimp to anxious retailers. His trucks were equipped with overhead spray systems that shot aerated water into live-bait compartments. Carrying twenty to forty thousand pink shrimp, his drivers prayed that they wouldn't break down as their heavily laden trucks sped down the highway.

Before recreational take limits were placed on catches, in both the size and number of trout, sportsmen loaded their ice chests with speckled trout. Customers lined up at the bait shops to buy live hoppers. Workers dipped and counted jumping, flouncing shrimp out of their vats and put them in buckets. They sold by the dozen and sometimes by the hundred until they were cleaned out. The next day, if the fish were still running, the bait shrimpers, truckers, and fishermen did it all over again.

Even with brisk business and high markups, it was an expensive and often unprofitable enterprise for the bait store owners. Pumps, filters, and aerators are costly and tend to break down on a broiling-hot day right after the dealer bought ten thousand shrimp from the bait truck at a dime apiece. Then it becomes a disaster: oxygen in the tank gets depleted, the weaker shrimp die and decompose, and the vats fill with foam. In a few hours, the rest of the shrimp die like flies. The salt air eats away at pumps and air conditioners, which sometimes have to be replaced two or three times a year. Backup pumps and aerators are a necessity.

Frozen shrimp in bait cups are second-rate. The chance of catching a fish improves dramatically with a live pink shrimp out of the bait bucket. When trout, redfish, or whiting smell and feel the microvolts of current generated by the shrimp's contracting muscles, they can't resist it. Almost every creature in the sea eats shrimp or tries to. There's probably not a fish or crab out there that hasn't eaten shrimp at some stage of its life cycle.

Tommy Robinson, a well-known tarpon fishing guide in Apalachicola, said, "When you fish for tarpon, bonefish, and snook, you use only the freshest, live shrimp to chum them in. If you use frozen or iced ones, all you'll catch are bonnethead sharks and catfish. While tarpon have large olfactory lobes, and hunt by smell as well as sight, their senses seem to be focused on a narrow range of smells. When a shrimp is still alive, the energy it emits rings the dinner bell."

Landings have gone up and down in the past ten years like a roller coaster, but in recent years the Florida catch has been nearly two million live bait shrimp, with a value of seven and a half million dollars. Ten years ago, roughly two million shrimp were caught, but the prices were almost half of what they are today. And around the nation, in Texas, Louisiana, and Alabama, fishermen are complaining that they can't afford to keep buying them, but they do anyway.

Ever since the roller-frame net was invented, landowners and visitors have complained that bait-shrimp shrimpers rip up the sea grass, causing it to wash up dead on the beach, but little evidence supports their claims. Shedding of sea grass leaves is a natural process, and great masses of sea grass leaves wash up on the beaches with or without bait shrimp boats working in them. Marine grasses are strongly rooted to successfully withstand the erosive effects of waves, tide, and currents. Scientists reported that the grass blades in heavily bait-trawled areas tended to be greener and freer from silt and encrusting algae than in nontrawled bottoms. Over the past seventy years, the roller frames have rolled over the same areas, catching untold millions of pink shrimp out of the living grass meadows, and the grass beds remain.

Some coastal dwellers hire beach-cleaning crews to remove washed-up grass from their sandy beaches, because it's unpleasant to walk on. However, the decaying sea grass is an essential component of the food chain, and removing it robs the sea of essential nutrients. As the decaying grass on the beach turns brown, it seethes with tiny beach hoppers or terrestrial amphipods that bounce like popcorn when disturbed. Churned by the tides and ground up by the beach sand, over time the grass breaks down into detritus particles and forms the very basis of the food chain. As the protein- and carbohydrate-rich particles get smaller and smaller, they're eaten by every detritovore on the bottom, including shrimp. Their nutrients and vitamins fertilize phytoplankton and diatoms, which in turn feed larval penaeid shrimp entering the bays from offshore, which will grow up to become bait shrimp, and later move offshore and become food shrimp.

Some recreational fishermen catch their own shrimp for either bait or eating, using cast nets. Attach the rope to one hand, coil the net in the other, fold it over a couple of times, and then toss it in the air, spinning it like a discus thrower. The net explodes into a giant monofilament fireworks display and comes crashing down into the water. After it sinks for a few seconds, grab the rope, and haul it in. The lead line contracts, entrapping the fish, crabs, and shrimp under the net so that they can't get out. Large-meshed cast nets are used to catch big fish to eat, and small-meshed nets are used to catch bait. The bait can be small fish, anchovies, silversides, and herring, or shrimp, all to be dangled in front of snook, trout, or tarpon.

Cast netting is one of the oldest forms of fishing, having been around for thousands of years. The nets have been found in Egyptian tombs, and the Bible contains references to casting nets. The Chinese used them, and so did Africans and many other cultures. On the same shores where people fish for pink shrimp in South Florida, the Calusa, one of the most powerful tribes encountered by the Spanish, threw their cast nets over giant schools of mullet.

In South Carolina, as many as 15,000 recreational fishermen buy licenses to cast net for white shrimp over a thirty-day period each fall. They are allowed to put out twenty poles each to mark the areas where the shrimp are concentrated, place cakes of fish meal and clay on the bottom, and throw over the bait. South Carolina white shrimp are olfactory animals that are attracted to the mix of clay, fish meal, and sometimes rabbit pellets.

Although there are a few places where fishermen can cast from shore or bridges, most go out in little boats and look for a deep hole in the winding creeks and bays where the shrimp bury themselves in the mud. At the beginning of the season, they may catch three or four to the cast, and now and then get a pound. It has more to do with the catcher than the bait. But as the season advances, the catch increases. When it does, the fishermen call their buddies on a cell phone to join them, and before long, thirty to fifty small boats surround a hole in a given area.

Casting over bait has become so popular over the past fifty years that South Carolina was forced to establish seasons and set limits—not so much to protect the resource, but to keep the fishermen from killing each other. When the shrimp are really running, and the holes become overcrowded, tempers flare as people argue that they were there first and their territories are being impinged upon. Before regulations were put in, the greedy put out two hundred poles, and when their markers got too close to each other, men fought and sunk each others' boats.

When someone did get killed, the State of South Carolina stepped in to regulate the fishery before the conflict escalated into open warfare. Politicians proposed a law that would allow recreational fishermen to cast net over bait but have no more than ten poles to an individual, spaced at least 60 feet apart.

Commercial shrimpers, who owned large seagoing vessels, protested the proposed law in front of the Capital, carrying picket signs, saying it was unfair that recreational baiters could cast nets in areas that were closed to them. The recreational fishermen were defeating the purpose of protecting the salt marsh nursery habitat where the shrimp grew up and were catching all the white shrimp before they could move out into the ocean.

The South Carolina Fish and Wildlife agency studied the conflict intensively, comparing landings over the years to see if they declined with the increase in recreational fishing. Despite the commercial shrimpers' perceptions, the landings did not decline. Even though thousands of people were baiting shrimp, filling cooler after cooler, they did not seem to make a dent in the offshore population. Shrimpers caught the same amount as they had before.

In the end, the shrimpers were unable to stop the law from going through. The sportsmen outnumbered the commercial fishermen ten to one, so they got what they wanted and won in the end. People who make their living from the sea are always at a disadvantage, since they are out there fishing for a living and don't have the time or social ability to fight political battles.

5

Microscopic Monsters

Shrimp are amorous creatures that have been making more shrimp since the beginning of time. Born offshore, planktonic baby penaeid shrimp move back to the coast in the first weeks of life, pouring into the inlets between barrier islands on incoming tides. Tiny pink and brown shrimp can be found in depths of an inch or two as a sheet of tidal water flows across the tide flats of the bays, sounds, and estuaries. Nearly microscopic white shrimp ride the currents all the way to the headwaters of the creeks, settling to the bottom in nearly freshwater, where predators are few. After several months of rapid growth in the nursery ground, the shrimp move into deeper creeks, rivers, and bays and then go offshore. There they mate, and the cycle is repeated (see Figure 5.1).

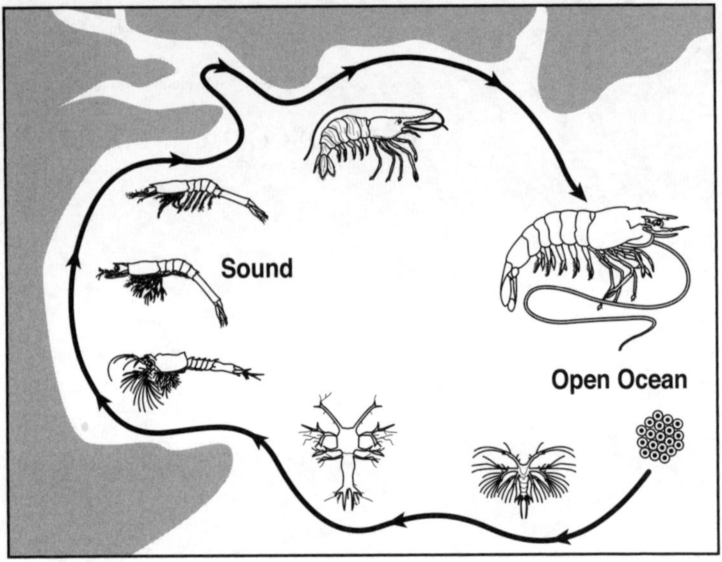

Figure 5.1 Shrimp migration

Illustration by Charlotte Ingram, courtesy of Georgia Sea Grant

After mating at dusk, a mother shrimp releases up to a million eggs. Within 24 hours, these eggs hatch into microscopic nauplius larvae. They molt, change, and grow in a world of eat-and-be-eaten, growing nearly an inch per month.

Like most aspects of shrimp behavior, courtship rituals have not been observed in the murky, dark waters of the wild—only in tanks on shrimp farms. Manuel Espejo of American Shrimp, Ltd. learned how to raise shrimp in Cuba before immigrating to the United States. He invited us to his hatchery in Panacea, Florida to watch them spawn after dark on the full moon.

Leading us to the closed-up six-foot-high black plastic tank, he said, "These are wild pink shrimp, and the mother shrimp are so nervous that if I stand beside the tank during the night, they can feel my presence and won't mate or lay eggs." Nevertheless, he allowed us a quick peek. Lifting the lid, we caught a glimpse of a jumbo female pink shrimp swimming off the bottom, going round and round and round, and two males sitting on the bottom. When the males of all species of penaeid shrimp are ready to mate, two bright white pearly balls called spermatophores appear on the underside of their abdomens, just behind their last pair of legs. When a ripe female broadcasts pheromones, all the males come running and begin circling her, trying to chase off other males.

By 3 a.m. the blessed event took place, and the tank was milky with a half million eggs. Under the microscope, a plethora of little circles with embryos inside came into focus. By the next morning they hatched into nauplius larvae. The technicians carefully screened them through a mesh woven tighter than silk and transferred them to a thousand-gallon maturation tank.

Hatcheries do everything they can to hasten spawning, feeding, and brood stock production. The hatchery shrimp have no worries about predatory plankton-eating fish or about getting enough to eat. A day later they had morphed into more developed protozoal larvae. Manny switched their diets first to microscopic algae and then to live brine shrimp, *Artemia*, which grow in landlocked briny lakes in Utah and California. The larvae were raised through four morphological stages and then were sold for a dime apiece to shrimp farmers in South Florida. Counting each shrimp in every batch was impossible, so the hatchery workers simply scooped them into a container that held a known quantity of shrimp.

Males are usually much smaller and fewer in number than the females, but they are deft lovers. A typical gravid lady shrimp weighs 45 grams, but her suitor weighs only 29 to 32 grams. The females need a bigger frame and more space to accommodate their eggs. What the males lack in size and number, they make up for in vigor.

Reproduction begins with a lot of touching and probing. The male gently approaches his mate and tenderly wraps his antennae over her carapace, delicately and persistently tapping her all over and caressing her. He pokes the female genital opening, called the thelycum, located just behind her last pair of legs, with his sharp pointed head spine, called a rostrum, and nudges her to get her off the bottom. A male pink or brown shrimp must wait until his mate molts and casts off her paper-thin shell before mating. After lying on her side and shaking off her molt, the body of a pink or brown shrimp is as soft as jelly. Like all freshly molted crustaceans, the bride is exceedingly fragile and vulnerable at this stage. If handled, no matter how carefully, her limbs and antennae would fall off. Shrimp farmers know not to handle them in that state. But a male pink or brown shrimp can hold his mate in a tight embrace and never ruffle a single one of her hairlike setae.

When sufficiently aroused, the females of the grooved pinks and browns leave the bottom and swim upward in a dizzying spiral. With the big female on top and the smaller male clinging beneath her, they dance together in the water column, going round and round until suddenly the female stops swimming and holds still, signaling to the male that she's ready to mate. He quickly swims under her and embraces her from the underside, pressing belly to belly, grasping her in his multitude of legs and wrapping his antennae around her.

Curving his body, turning crosswise, he pastes his sperm packet onto her slit. They hang in the water column for several minutes, looking like the letter T as they slowly sink in the water column. No doubt hanging there together in mid-water, even for those few brief moments, is a dangerous time for copulating shrimp. It seems likely that the pair would make a tasty morsel for any passing jack, tarpon, or shark, yet many succeed. How they avoid being eaten isn't understood.

When mating is finished, the couple break apart, drop to the bottom, and swim off on their separate ways. She can keep his seed alive for weeks or months, until the time is right for her to release her eggs. A female pink or brown shrimp can fertilize three batches of eggs on the same sperm packet.

Female white shrimp cannot store sperm as the other species can. They must release their eggs into the ocean immediately after mating. White shrimp carry quick sex to an extreme. In a high-speed collision, the male white plasters the underside of the female's shell with his sperm packet and swims on. She doesn't have to molt; her shell remains hard. The fact that white shrimp, unlike pink shrimp, don't have to survive a molt to mate is one of the main reasons white shrimp are the preferred farmed shrimp. Their quick, external fertilization, plus their ability to live in crowded ponds, make white shrimp better animals for aquaculture than pink or brown shrimp. Most of the shrimp we eat in restaurants and supermarkets nowadays are one or two species of whites that come from farms in Asia and South America, not the open sea.

Wild penaeid shrimp release their eggs offshore in deep, high-salinity waters with stable temperatures. The onset of spawning usually is associated with an increase in water temperature in late March and April, and it extends through the summer until August or September. Most temperate-zone marine animals spawn in the early spring, when nourishing microscopic plant cells called phytoplankton are abundant. If the mother shrimp discharged her eggs in the dead of winter, when the phytoplankton is less, most of her offspring would starve to death.

When it's time to release her fertilized eggs, the mother shrimp swims high up into the water column in another spiraling dance. She rips into the spermatophore packet using her last set of feet, designed for that purpose. In a series of violent contractions, the mother forces her eggs from her body, discharging them through ducts at the base of her third leg. As they pass through the spermatophore, she fans her pleopods, mixing them with sperm and fertilizing them as they're discharged into the sea.

Her spherical floating eggs measure a quarter of a millimeter in diameter. A female shrimp produces between half a million and a million eggs during a single spawning. She gives no maternal care. As broadcast spawners, they release as many eggs as possible into the environment in the hopes that a few will survive. And those that do, at least for a little while, float on. As long as the currents flow, eggs laid by invertebrates and fishes may travel for days before hatching.

When her dance is over, the female dives to the safety of the bottom, and the eggs that are left inside are reabsorbed into her body. Out of millions of reproductive adults, only 20 percent of gravid females may spawn on a given night, but they produce a great rain of eggs. Most of the eggs float above the bottom, held afloat by a glob of oil in their yolk that gives them a slightly positive buoyancy and prevents them from sinking.

The developing shrimp eggs join a plethora of eggs from starfish, urchins, clams, snails, and other species drifting along in the plankton. After fertilization, the shrimp egg cells divide rapidly until they hatch, bursting from their cramped spheres with convulsive movements to hang in the water.

Barely a third of a millimeter at birth, with only three sets of appendages, nauplius larvae look like tiny transparent spiders, more like a ball than a shrimp. Feeble at first, they begin kicking, joining in the fray of the living. This is the first of many microscopic larval stages before they change into their final adult form. The first act of a newly hatched shrimp is to remove the mucous and bits of embryonic debris that cling to it. It then drifts for a few seconds at rest.

If they survive the next several weeks, they develop offshore through a series of naupliar, protozoal, and mysis stages before metamorphosing into post-larvae and then into tiny juvenile shrimp as they arrive in the estuaries. Each stage is bigger, stronger, and has more sensory organs than the previous stage. With each successive molt, or shedding of their exoskeleton, they gain more hairs and spines on their bodies that keep them from sinking, as well as more sensory organs and appendages. The juvenile shrimp is as different from the early larval stages as a butterfly is from a caterpillar.

Nauplius larvae do not eat. Protozoans whirl around them, and bacteria float in the sea, but the larvae pay them no mind. The yolk sacks attached to their bodies contain enough rich protein and lipids for energy to keep them alive for 24 to 36 hours. After burning through the food, they separate from their spent sacks like a space shuttle jettisoning its empty fuel tanks. The most prominent feature of any crustacean nauplius larva is the single eye spot on the top of the head, which enables it to orient to the light above. In the laboratory larvae are drawn to overhead lights. Simply shine a flashlight on the side of the beaker, and the little specks come to life and head for the beam with explosive, convulsive movements. Out at sea, sunlight and moonlight shining on the surface draws them.

After going through a series of nauplius stages, they molt into the first of several protozoal stages. The exact number of stages varies from species to species. They acquire a gut and the tiny tails that may someday grow to be food for some lucky person. They can now grab the green orbs of phytoplankton that orbit them, rip through their membranes, and eat their first meal. Hours later, they molt again, augmenting their diet with meat. With more limbs and extended hairs, the protozoal shrimp can right themselves, swim up and down in the water column, and shoot forward.

With wriggling spines, segments, mouthparts, and twitching antennae, the shrimplets charge forward and snap up the whirling copepods that orbit them. The microshrimp engulfs its prey in a net of legs, holding it firmly so that it cannot escape, and pulls it into a terrible maw. With a sideways, scissors-like action, the mandibles, which look a little like the serrated pinchers of blue crabs, dismember their struggling prey.

No longer sitting there helplessly, waiting to be eaten, the protozoal larvae can now dodge predators as they seize copepods, arrow worms, and fish larvae in a blinding flash. Now a full millimeter in length, they look less like spiders and more like shrimp; their newly formed bulbous eyes stick out at right angles from the side of their heads like a hammerhead shark. Chemoreceptive organs are now fully tuned and functional, and the antennae sense the presence of life about them.

Incessantly bobbing, taking little bows like a performer on stage, the minuscule transparent protozoal shrimp use their new tails to wriggle away from the tiny hungry larval squid that change color as they shoot out their tentacles, trying to ensnare the shrimp. Crab larvae extend their newly acquired claws and bear down on the shrimp, but unlike the earlier helpless nauplius larvae, the protozoals can escape. Their ball-like bodies have stretched out and become spiny and elongated, making them harder for predators like glass minnows to swallow.

Growth happens fast in a sun-warmed sea. Always shrimp move to the light, to the surface waters where there are fewer small fishes to eat them. At the surface the plant cells become so dense, they turn the waters of the continental shelf from blue to a greenish gray color, cloudy and turbid with phytoplankton. Green and brown stains can be seen from satellites flying 40 miles overhead along the continental shelves where shrimp spawn. According to NASA, 90 percent of the oxygen on Earth comes from phytoplankton.

The sun beats down on the sea, and warmth penetrates into the chilly water, stimulating the growth of microscopic plants. As the phytoplankton increase with the lengthening daylight and warming temperatures, so do the minute and microscopic zooplankton animals that feed on them. A single cup of seawater contains millions of microscopic orbs and chains of plant cells that are being fed upon by great numbers of microscopic animals. To envision the density of plankton, look at the fine suspension of dust in a sunbeam—minute particles floating in space.

Flowing along with the protozoal shrimp is a vast array of other growing invertebrate larvae beating their feeble appendages, or fish fry undulating their fins. Embedded in a vast matrix of planktonic life, the shrimp move along with the triangular and rocket-shaped young of starfish. Microscopic sea urchins, and sea cucumbers, clams, crabs, and barnacles look nothing like their parents. Baby snails that will spend their entire adulthood crawling in the mud swim with octopus-like arms to keep afloat. Snails yet to grow shells and worms the size of dust particles swim beside the transparent wraiths of baby mackerel, bluefish, and snapper spawned in the waters above the continental shelf, all growing, changing shape, and eating each other.

Clouds of microscopic larvae from many species of crustaceans drift together, each separate from other species. Perhaps they are brothers and sisters, spawned from common parents, or their aggregation is simply the result of density sorting and physics. Microcrabs, worms, sea urchins, and sand dollars, freshly hatched from the egg are swept around, fed, and dispersed by the coastal currents. With them are tiny larval fish with big eyes and bodies so transparent they hardly cast a shadow. Together they make up a whole ecosystem, a food chain filled with interrelationships we cannot see and can barely measure. For all the bizarre babies, survival is a chancy business and a matter of luck. It fails some years, and babies die by the billions because the weather is either too wet, too dry, too cold, or too hot, or the wind blows in the wrong direction at the wrong time. It creates cycles of boom and bust not only for shrimp and shrimpers but for almost all smaller marine species.

As the weather warms and plankton booms, the surface waters boil with schools of bait fish, silvery pilchards, sardines, menhaden, and anchovies stuffing themselves on shrimp larvae and eggs as well as all the other plankton. Woe unto any larval shrimp if they are happened upon by great clouds of voracious plankton-eating anchovies. Nothing but their sheer numbers will save them, because predators eat until they can eat no more. Amazingly, spawning millions of young, most of which will be quickly eaten, is a successful reproductive strategy that is used by most marine invertebrate and fish species. The frenzy of little fish often attracts sharks, jacks, tarpon, and cobia, who become the plankton's protectors. They pass by the specks of life unnoticed and smash into the schooling balls of silvery plankton-eating bait fish. And while sharks and jacks are attacking frantic pilchard, menhaden, and anchovies from below, flocks of seabirds dive on them from above. Predator and prey float together, tasting the inshore waters in a world of eat-and-be-eaten.

Carried along by surface currents they are pushed shoreward by favorable winds over bottoms that become ever more shallow and brackish. Storms flush giant plumes of dark freshwater out to sea. As the developing protozoal larvae move from the clear blue offshore waters into the turbid green waters that cover the continental shelf, they taste the "swamp tea" from the rivers flowing out to sea. Shrimp begin eating particles of detritus, leaves, and bits of organic matter from sea grasses and upland vegetation as they approach the coast, along with their diet of phytoplankton and zooplankton.

Laboratory studies indicate that shrimp larvae gravitate toward lower salinity, always moving toward coastal waters. It takes the right combination of not too dry and not too wet to grow shrimp. Estuarine marine animals are adapted to a gradual mixing of salt water and freshwater and to gradual changes as salinity rises in droughts and declines in rainy seasons. If it's a wet season, the migrating crustaceans don't have to go far to meet the brackish waters. They spread out and expand their territory into the rich floodplains and forage on the decaying leaf litter, twigs, seeds, and insects. But in dry years, they may go for miles and miles inland before they reach freshwater and escape their predators and competitors. In periods of extreme drought, that might be at the very headwaters of the river. Pickings are lean there, because their foraging grounds are vastly reduced. If salinity in the bays and rivers becomes the same strength as the open sea, the shrimp grow slowly, or not at all. Shrimp larvae in uncountable numbers die unnoticed. "Shrimp are like grass. If it rains, they grow, and if it's dry, they don't," commented one shrimper.

When rain is scarce, shrimpers go broke burning fuel and hauling up empty nets, and finally they tie up their boats in disgust and look for other work. Recreational fishermen get tired of catching nothing, and tourists go home. Restaurants and motels suffer; the coastal economy takes a dive, and everyone prays for rain.

The shrimp larvae that have survived their planktonic predators have undergone numerous molts and changed from helpless spiders to tiny predatory wraiths. Yet the game is far from over. The protozoal shrimp molt again and achieve the first of several mysis stages. Mysis-staged penaeid shrimp are not to be confused with the tiny mysid or opossum shrimp that never become large and keep that name throughout their lifetimes. Now looking like a true but very tiny shrimp, and having grown to a whopping quarter of an inch, they no longer are microscopic and can be seen by hungry fish. The water column that once afforded protection and abundant food becomes a place of danger. They abandon their planktonic existence, beat their way down to the seafloor, and dive into the mud for protection.

Even though they're stronger and faster, the molt from a protozoal stage three to a mysis stage one is stressful. It is a time of crisis and vulnerability in both the ocean and the hatchery. The mysis-stage shrimplets have now moved into inshore waters. If the salinity drops from hard rains, or it's too dry and the estuaries take on the same salinity as the open sea, only 40 percent of the year's crop may survive. A hard freeze in the spring can kill the entire year's crop as seawater temperatures become too cold for them to survive. In this world of perpetual munchies, the mortality of larval shrimp is enormous. Fisheries biologists know there are good years and bad years for shrimp, and years in between, and there's nothing that man, with all his fisheries laws and regulations, can do to change that. In the end, most of the larvae produced from the original half million to a million eggs that were spawned a week ago will get eaten. Still, it's a very effective strategy, because shrimp have been around for more than 450 million years.

While survival is chancy for baby shrimp at sea, in a hatchery, where temperature is controlled and technicians disburse food constantly, the next two mysis stages begin two to three days after hatching. However, out at sea, it may take a week or two, depending on water temperature and food supply. The total time for larval development takes approximately 17 days at a temperature of 24 degrees Celsius.

Before the mysis stage, shrimp are swept through the island passes into the brackish bays and sounds, they molt again and become post-larvae. Several post-larval stages occur before metamorphosis into the juvenile stage. By the time they enter the bays and sounds, they have grown from a third of a millimeter to an enormous 8 to 10 millimeters (nearly half an inch) from the top of their pointed heads to the tip of their tails. The biggest influxes of post-larvae come during full and new moons, when tidal currents are strongest.

Pinks, browns, or whites, they arrive in the estuaries as fully formed little shrimp; they have grown all nineteen pairs of appendages, swimmerets, antennae, and a pair of mandibles. Post-larval shrimp drop to the bottom like swarms of locusts, snatching worms from their burrows as soon as they hit bottom. Little shrimp eat nematodes and little writhing worms; big shrimp attack fat bristleworms, scaleworms, and lugworms. They grab and run with them, always on the move. For an adult female prawn to reproduce, she must have high-energy worm flesh in abundance, or she won't be able to make eggs.

Post-larval shrimp develop directly into juveniles and grow rapidly, up to two and a half inches a month (see Figure 5.2). They remain in the bays, growing rapidly until they exhaust their food supply and have to leave. By that time, they've outgrown many of their smaller predators. At four and a half inches, they are ready to leave and head to sea.

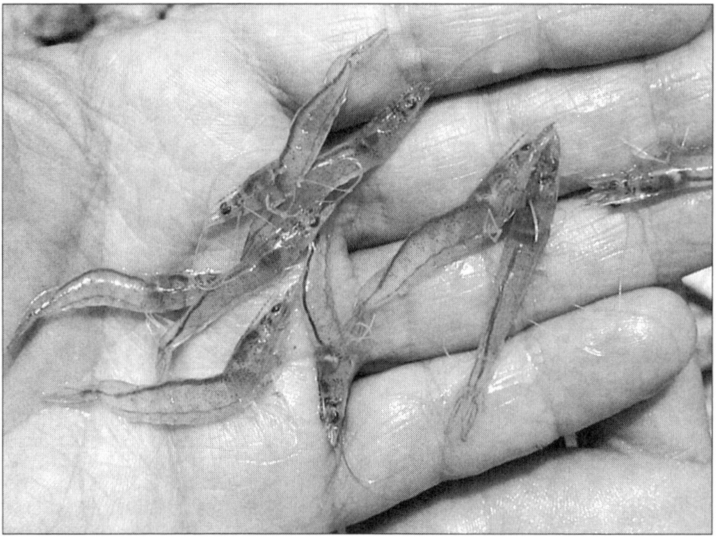

Figure 5.2 Juvenile white shrimp

Courtesy of American Shrimp, LLC

While juvenile pink shrimp settle into the sea grass meadows, white and brown shrimp thrive on shallow, muddy, and peaty bottoms high in organic detritus. Needing an abundance of marsh grasses or mangroves with moderate to low salinities, they advance up the creeks to the head of the bays and mangrove thickets, getting new food from the detritus that continually forms. White shrimp go up the creeks into freshwater, which is a barrier to most marine fish.

The arriving transparent juvenile shrimp flooding into the bays and estuaries are a feast for the assembled gauntlet of fish. Night or day, all shrimps' lives are filled with danger as ambush predators lie in wait. Shrimp make up the better part of the food base for many estuarine and bay fishes, including juvenile snapper and other recreationally important game fish.

As the young shrimp spread out over the mud bottoms, toadfish hidden beneath sunken logs or rocks inhale their quarry with lightning speed. Small-mouthed juvenile flounders are especially fond of post-larval and juvenile shrimp. Buried in the substrate, the flounder looks forever upward with eyes on top of its head, waiting to surprise hapless shrimp, crab, or killifish as they come by. Then it explodes from the sand, grabbing the shrimp in its powerful jaws, and swallows it in a millisecond. Juvenile redfish, trout, whiting, skipjack, sharks, stingrays, tarpon—not many fish won't eat shrimp. Spotted sea trout eat shrimp almost exclusively.

Around the world, shrimp of all species are the foundation of the food chain. In cold oceans, krill, tiny relatives of shrimp, swarm by the trillions often turning the water blood red and are fed upon by whales, seals, sea lions, and factory ships that process them into meal and paste. People also feed their expensive and finicky salt-water aquarium fish live grass shrimp and watch their pets go nuts trying to catch them. Even the impossible-to-feed round batfish, which would rather starve to death than eat dead food, will hop along the bottom of an aquarium, snapping up live feeder shrimp.

But penaeid shrimp have their defense, mostly avoidance. Masters of escape and gymnastics, shrimp are the acrobats of the sea. They are hydro-dynamically designed escape artists with muscular abdomens to jet backwards using their tail fins as rudders and spoilers to turn on a dime. A millisecond before a speckled trout's jaws snap down on it, the shrimp explodes, turns somersaults, and jackknifes. It can dart backward at a rate of one meter or more in less than a second. In shallow marshes, white shrimp often rocket out of the water, skip, jump, and get away. They skip over the water, beat their swimmerets to speed forward, or fan themselves into the mud until even their long whiskers are buried. They spend their entire lives looking around for predators and watching for food with bulbous eyes that can see 360 degrees on eyestalks.

They also have armor. The formidable serrated rostrum or spine that projects from the penaeid shrimp's head improves its hydrodynamics by acting like the keel of a boat, but it also protects the eyes from attack and is used in the shrimps' mating rituals. Tiny grass shrimp, *Palemonetes*, have been observed in an aquarium standing their ground against marsh minnows that are four times their size, raised on their slender legs, jabbing and thrusting their rostrums at the attackers until they drive away the fish. As the rostrum grows, it gets larger in proportion to the shrimp's body. Small shrimp, which average three to four inches in length, have relatively smaller head spines than the mature, powerful jumbos found offshore.

Along the Gulf Coast and South Atlantic, an adolescent shrimp can grow an inch a month or more. A good-sized shrimp is around seven inches. Females achieve a larger size than males and reach reproductive maturity at five and a half inches. At that size they are no longer "all eyeballs and whiskers," as the shrimpers say.

At the end of the long summer's growing season, when cold fronts come, pinks leave the sea grasses, and juvenile white and brown shrimp leave the protection of the marshes, rivers, and creeks. They swarm by the millions into the bays and then head for open waters, causing predatory fish to go into a feeding frenzy and grab everything in sight. When that happens, lucky recreational fishermen fill their coolers with red drum, sea trout, and grey snapper. Adult browns migrate to briny deep waters off Texas and Louisiana and are found at depths between 150 and 300 feet. Adult whites stay inshore in shallow water, in sight of the barrier islands. One tagged white shrimp remarkably traveled 360 miles along the coast in ninety-five days.

"White shrimp will run you crazy," one shrimper said. "They're here one minute and gone the next. When they get ready to go, their legs turn red, and they take off. You can spend your whole life chasing them up and down the beach, run aground on a shoal, and beat the bottom out of your boat. That's why lots of captains won't fool with them unless they have to. Me, I'd rather drag for hoppers and brownies!"

Flocks of screaming seagulls hover; pelicans and cormorants dive on white shrimp; and schools of redfish, whiting, weakfish, and flounder follow in their wake, snapping them up. Converting detritus into protein, gaining mass and length, they move slowly out of the highly brackish water into the more saline open bays, gaining a mile or two a week. They generally appear on the fishing grounds in June or July, depending on water temperature, salinity, and food supply. When they do, the biggest and most efficient predator of all awaits them. Their sheer numbers overwhelm and satiate all their predators but one: the shrimp boat. Most live one or two years, dodging the nets, completing the cycle of life.

For penaeid shrimp, winning in the game of life means growing up, migrating out of the sheltered bays into open waters, and broadcasting trillions upon trillions of eggs so that the next generation can begin anew. If a few million or billion shrimp are lost, so be it. In biology it's the species as a whole that counts, not the individual. Shrimp have been winning the population game for millions of years by saturating the sea with their spawn in estuaries and continental shelf bottoms throughout the world.

6

The Shrimp Run

With the sea rusting away the rigging and eating up the bottom of the boat, and the cost of fuel never going down, shrimping is a rough way to make a living. A shrimp fisherman must fight the weather, run from hurricanes, and watch for sharks tearing holes in his nets.

It's especially rough making a living shrimping in the winter. In the northern Gulf of Mexico, cold front after cold front sweeps down from Canada, the plankton die, and the water is so clear and lifeless that hardly a worm crawls. The sea becomes so icy cold that shrimp vanish into the mud. Even the big trawlers, fishing in a hundred feet of water where it's a little warmer, can't catch enough shrimp to pay expenses. Fishermen call it scrapping.

Shrimpers are optimistic; they have to be, or they wouldn't be shrimpers. Everyone waits for the weather to warm, living on hope, and everyone goes broke until it does. Between getting their boats ready for the fishing season, fishermen hang around the coffee shop, exchange gossip, tease the waitresses, complain about rising fuel prices, and talk optimistically about the last run and the lucky guy who loaded down his boat late in February three years ago, when it was least expected. They never quit talking about the big pink shrimp or hopper run that unpredictably occurs in the early spring of some years but not all. In the 1970s, in what the shrimpers now call "the glory days," it was hard work. But there was money to be made, and the men and women who worked the sea and caught and processed the shrimp felt good about their work (see Figure 6.1).

Figure 6.1 Shrimp boats at dock in Key West, awaiting the run
Courtesy of the State Archives of Florida

Captain Edward Keith was a good friend. He worked on our little shrimp boat in the late '60s when he was fourteen. He learned all the scientific names of the animals and how to keep them alive. Later he became a shrimp captain, running boats for other people, until he finally bought one of his own. When the National Cancer Institute needed a ton of bryozoans, an obscure moss animal that contained bryostatin, a crystalline substance that proved to have powerful cancer-preventing properties, he and his brothers rose to the challenge and dragged up a ton. It was Edward Keith who helped us supply the Harvard Medical School with live electric rays for research into neurodegenerative diseases like Alzheimer's. He brought in specimens of Kemp's ridley sea turtles and helped us tag and release them. Many people in the town of Panacea depended on him at one time or another, whether it was rescuing someone from a storm, or providing the restaurant with fresh seafood, or helping fix your car.

One cold morning in the Coastal Cafe we sat drinking coffee and talking with Captain Edward about the great shrimp run. In the days before the industry was choked with regulations and foreign competition, shrimping was the American Dream. Trawlers unloaded their catches at fish houses, and money flowed like water. In a few years a man could go from deckhand to captain, captain to boat owner, boat owner to fleet owner, and fleet owner to seafood processor. Fishermen mortgaged their houses to purchase boats, buy fuel, make repairs, and get ready for the run. Some did well; a few got rich; many went bankrupt.

"I'll never forget when that run started," he told us. "It was dark, just another night. We dropped our nets out, and they came up exploding with pink shrimp. Did our spirits brighten! You can tell almost the instant the big run starts. Suddenly you're catching shrimp that look clear and have bright red legs. There's so many that they cover the sea bottoms for miles and miles. In minutes it was all over the radio."

While no one understands the run or why the shrimp do it, when north Florida hoppers and south Florida pinks go, they get so thick that some swim on top of the water, usually with an accompanying armada of sea robins, small fish that usually stay on the bottom. Flounder and trout come up in the nets with shrimp whiskers hanging out of their mouths. When a run of shrimp starts, there's very little bycatch. The shrimp swarm over the bottom, with whiskers touching tails. Predators get out of the way, as the shrimp seem to overwhelm and drive out everything else.

"Soon as the deck was cleared off, and the shrimp were shoveled down below, the nets came up with another load. It was like coal pouring down a chute. Shrimp rained out on the deck, and for the next three hours we all shoveled clean shrimp down into the hold, maybe picking out a fish or two, but mostly it was a clean catch," Edward continued as he finished his coffee.

When the sun sets, the nocturnal pink shrimp kick themselves out of the sand and feed along the bottom. When the first rays of morning sunlight penetrate into the night sea, the shrimp dig their way into the mud and rest until dark. Buried five inches deep, with not even their whiskers sticking out, they are safe from predators and shrimp nets. But not during the run. Then hoppers keep moving day and night, and so do the trawlers above them. Shrimpers say the shrimp get so thick that there's no room for them on the bottom, so they have to keep moving. Hundreds, sometimes thousands of pounds are caught in a single one-hour tow. Boats drag round the clock from night until morning, continue all day long, and into the night.

"By the third day we walked around with our eyes glassed over. When the shrimp are running, we don't care about the weather," Edward recalled happily. "We work with rain beating in our faces and when it's so rough you can't hardly stand up, 'cause that's when we make our money. If your nets get tangled or something goes wrong, you can lose thousands of dollars in an hour."

One year, the run went on for a solid week. Captains struggled to stay awake and run the boat. That's when accidents happen—sometimes terrible ones. Shrimping is a dangerous profession, going out in rough seas, with heavy gear and winches. Deckhands have fallen overboard, drowned, been eaten by sharks, or had their brains knocked out by blocks falling on their heads. Cables and winches can cut off a sleepy man's arm, or worse. One year in Louisiana a man fell asleep at the wheel and crashed into an oil platform, and the boat burst into flames. Every year someone runs aground, and pounding waves on the shoals beat the bottoms out of the boat.

Johnny, Edward's brother and deckhand, sat at the table with us, wearing a greasy threadbare sweater. Their jeans were faded by salt, spray, sun, cold, and sea, not beaten up by machines to look fashionable. They wore woolen caps, and their hands were rough from working with the sea. They had contempt for "gravy" shrimpers who jumped in when the shrimp were running and then went back to other jobs. Those who do this sometimes-brutal work love shrimping and wouldn't do anything else—especially when the shrimp run and the money is good.

Edward Keith explained, "There are some captains who are deep-water men and others who hunt white shrimp almost in the waves on the beach. You put a man who wants to catch white shrimp out in that deep water hopper fishing, and he won't do worth a flip. Deep-water shrimpers hate working on the beach; they're scared of a shoal. You can get thrown up on a sandbar, and it will beat the bottom out. But when you have your own boat, there are no bosses to answer to—except the banks, the fuel dock, icehouse, boatyards, marine hardware stores, and mechanics!"

"It'll pay off if we get a good run," Johnny put in happily, "but it's hard work. My back was breaking crouching over that pile [of shrimp], my hands got sore and waterlogged, but I didn't care. I was making money—more than I ever made in my life!"

"The boat stocked a hundred thousand dollars worth of shrimp that run. Then, just as suddenly as it started," Edward recalled wistfully, "the run was over. The shrimp just vanished, and the nets came up empty. With more than a hundred boats working day and night, maybe we caught them all. But if that was the case," he puzzled, "there should have been a slower falling off, with fewer pounds caught every night."

If the shrimp were going to run, they seemed to do it when the water temperature warmed into the high 60s. But in spite of the millions of dollars that has been spent on research, tagging shrimp, studying their life cycles, and tracking their movements and migrations, the run remains a mystery. Where did they go? No one knew—not the shrimpers or the biologists. They don't even know when a run will happen. It could be in ten years, or it could be tomorrow.

As the weeks passed, the sun warmed the sea, and finally the weather radio reported that the seawater temperature offshore was 64 degrees. In early April, Captain Edward decided to give it a try, and we went along. He ran his friend's sixty-eight-foot boat, the *Lady Murle*, for a day and a half east from Apalachicola to Duer's Channel off Alligator Point, where the shrimp usually showed up first. We got there in late afternoon and went straight to work. Penaeid shrimp, lobsters, and many other forms of marine life are most active just after the sun sets and just before it rises (see Figure 6.2).

Figure 6.2 Hanging up nets on a boat before setting them out

Courtesy of the State Archives of Florida

"Why are you setting out now? It's still daylight. Pinks are buried."

"Yeah, except when they're not. There's always a chance of a surprise: a big sloppy pile of shrimp," Edward said optimistically. "You just never know. I've been shrimping twenty years now, and I still don't know what they do. Years back, I thought I did. I used to keep records, but I don't anymore. They always change."

The sun went down, the night sky became studded with stars, and the fishermen's spirits lifted as they dropped the big nets overboard. The heavy weighted doors splashed into the sea, dragging down the great sock-shaped net behind them. The cables spun off the winch, and the nets were soon 200 feet behind the boat.

Several hours later it was time for the big otter trawls to come up. The skipper slowed the boat and threw the winches into gear. In a few moments hundreds of feet of steel cable were wound onto the winch drums, and the otter doors rose from sea with a splash and dangled from the outriggers. Edward shoved down the throttle and the *Lady Murle* raced ahead, washing the sand and mud out of the catch and pushing it to the end of the net. His deckhand looped the lazy line, which connected the doors to the tail bag, around the revolving brass cathead, and hauled up the tail bag of the first of the two nets.

As the first great webbed sock was lifted out of the sea, gorged with a variety of creatures, Edward's brother Johnny snatched the release rope, and the bloated bag disgorged a small mountain of bottom life on deck. Horseshoe crabs lurched over the bodies of struggling fish, sponges, and starfish. The noise was incredible, with grunts grunting and croakers croaking and stingrays beating their wings, but the fishermen didn't hear it. They were focused on shrimp. They paid no attention to the speckled moray eel that slithered toward the scupper hole with its mouth agape. Nor did the shrimpers see the hundreds of small iridescent swimming crabs with long delicate pinchers that scurried sideways off the pile and over their boots. Johnny looked hopefully over the side to see how many fiery orange shrimp eyes were blazing up at him, but he saw in an instant that only a few pounds of shrimp were mixed in the catch.

Gloomily they hauled back the second net and opened it. It was equally bad, with only a few of those beautiful jumping pinks flipping their tails. The water was still too cold for them to appear in any number. Edward kicked his boot through the pile of squirming life. "We'll be lucky we get thirty pounds out of this!"

His prediction was accurate. An hour later, after Johnny had culled the catch, they had only half a basket of shrimp. A small bay boat would have been happy with that catch, but a big fuel-burning double rig would go broke at that rate. Still, the night was young; shrimp were fickle and could pop out of the mud at any time. They put the nets back over. "Them boys back home weren't kidding when they said they was catching a few shrimp," Edward grumbled. It was what the shrimpers called "scrapping." It had been five years since there was a real, honest-to-God run, not just a flurry of pink shrimp.

Hours passed. While Johnny shoveled trash overboard, Edward sat at the wheel, periodically flicking on the depth recorder to ensure that he was not about to run aground. He watched the radar that showed him where other boats were working and how close he was getting to the beaches. Now and then, he shone his spotlight on the water to make sure that he was not in the midst of crab traps.

Along the eastern Gulf of Mexico a chronic conflict existed between crab fishermen and shrimpers in the 1970s. When crabbers set their pots in prime shrimping areas and shrimpers caught a net full of the wire baskets, trouble started. Crabbers had shrimpers arrested and dragged into court for destroying their trap lines. Shrimpers claimed the crab fishermen were obstructing navigable waters. An uneasy truce eventually more or less worked itself out, with each group keeping to different areas, but every now and then a boat new to the area dragged into the traps, and the trouble started again. In recent decades, government regulators separated the two fisheries by limiting where each group could fish.

The hoppers didn't run that night, nor would they that season. Edward and Johnny caught six baskets of shrimp, which was barely enough to make expenses. In the morning we went back to the dock and unloaded the catch. As the days grew longer, the fleet dragged up and down the North Florida coast, working off Dog Island, into Carrabelle Sound, off St. George Island and St. Joe Bay over to Panama City and back again. Some boats took off and ran 200 miles down to Cedar Keys looking for shrimp. But there were few shrimp anywhere. They caught barely enough shrimp to make boat and mortgage payments, do some repairs, buy fuel, and get ready for Brownie Season.

Every year after the hoppers played out, North Florida shrimpers headed over to Mississippi to catch opening day of brown shrimp season. While pinks grow up in sea grass meadows, mud-loving brownies grow in low-salinity coastal salt marshes.

The great Mississippi River carries the rich eroding soils of America's farmlands down to the Gulf of Mexico, which makes the sea bloom with shrimp, crabs, fish, and other protein. The Mississippi coastline, which includes large bays at Bay Saint Louis, Biloxi, and Pascagoula, is separated from the Gulf of Mexico proper by the shallow Mississippi Sound, which is partially enclosed by the Petit Bois, Horn, Ship, and Cat Islands. Brown shrimp live primarily in the Mississippi and Texas drainage areas. They share the same habitat as white shrimp, but subtle differences in their behavior and their preference for different bottoms generally keep the two species apart. Brown and white shrimp can coexist because brownies are largely nocturnal, while white shrimp forage during the day, and they arrive in the estuaries at different seasons. Post-larval and juvenile white shrimp travel far up rivers to areas where the water is almost fresh, while browns remain in the deeper, saltier parts of the bays.

At the right season—spring for brown shrimp and fall for white shrimp—juvenile shrimp leave the protection of the marshes, rivers, and creeks of the central Gulf of Mexico, where the shrimping culture began. They swarm by the millions back through the bays, heading for open offshore waters in huge migrations. Flocks of screaming seagulls hover; pelicans and cormorants dive on them; and schools of redfish, whiting, weakfish, and flounder follow in their wake, snapping them up.

When the shrimp grow large enough to harvest, the Marine Resources Commission opens the bays to fishermen. It's a tightly managed fishery. Every year the Commission does a brisk business selling licenses that permit shrimping in the bays and sounds. Mindful of protecting its $30 million annual seafood industry, the sovereign State of Mississippi stands firm, keeping the bays closed until the shrimp are big enough. Every week, state biologists sample the marshes, creeks, and bays to see if the brownies have reached a minimum count of 65 shrimp to the pound. In a good year, with just the right salinity and food, the shrimp grow rapidly, and the bay usually opens in early June. But in years with a cold winter and heavy rains, which lower the salinity of the water and slow growth, the shrimp are small, and the bays stay closed for a few more weeks.

As the shrimp approach legal size, boats come from all over the Gulf and wait for the bays to open. A few become impatient, sneak out at night with their lights off, and load down their boats with undersized shrimp, but the penalties are severe. Getting caught with shrimp before the opening results in a stiff fine, and violators risk having their boat confiscated by wardens. Poachers are frowned upon by the other shrimpers. But when a vessel is confiscated and sold, there's an unwritten rule that if anyone other than the original owner bids on the boat and wins, he may find his newly acquired vessel sunk at the dock.

The wardens do their best to reduce temptation by walking the docks in their blue uniforms, wearing badges and guns. They check licenses, report on the biologists' survey results, and make a show of patrolling the bays in police boats. In past years, wardens didn't like arresting commercial fishermen. Up until the 1990s, many were ex-shrimpers, fishermen who gave up fishing in favor of a steady job. One said, "If you arrest a man trying to make a living, you're arresting his wife and kids. Being a fish cop isn't like being a deputy sheriff, where you get to throw some loudmouthed drunk driver in jail." Today, that relaxed attitude has largely disappeared as marine police are no longer drawn from the ranks of fishermen.

When fisheries biologists find that a pound of intact unheaded shrimp has sixty-eight or fewer shrimp, the season finally opens, and an epidemic of shrimp fever breaks out. The Pascagoula shipyards practically shut down as men call in sick. Recreational fishermen launch boats that had been sitting in the garage all winter, and they get out the twelve-foot trawls that they are legally allowed to pull under a recreational license. They cram their freezers with brownies for parties, the blessing of the fleet, and upcoming football games.

Hoping to see a brown shrimp run, I (Jack) joined the *Lady Murle* that year to catch opening day in Pascagoula. It's mostly a small bay boat fishery, but big trawlers like ours were allowed to drag in the bays and sounds if they used only one net. A few days after we arrived, the fisheries biologists announced that the shrimp were large enough to harvest. At twelve o'clock on the official opening day, sirens sounded across the Pascagoula Bay, and hundreds of otter trawl doors and nets splashed down. There were Louisiana-built Lafayette skiffs, with high flared bows, and a few picturesque, antiquated Mississippi bow luggers, designed after the old sailing schooners that fished the central Gulf in the 1800s. The pilothouse and living quarters were built aft, and the mast and boom were forward. The deck arrangement enabled the captain to keep an eye on his crew during drags, but at night it was problematic, because the deck lights blinded the captain. If there was fog or inclement weather, someone had to run up to the bow and look to see what was coming. The vessels were designed for dredging oysters and for white shrimping, which are daylight fisheries.

On opening day, boats ran over each other as tons of small shrimp were caught. The excitement grew as men towed hard, blowing black smoke out of the exhaust stacks of their diesel boats. Skiffs with outboard motors pulled beside double-rigged trawlers like ours that kept one net on deck and pulled and centered the other with a hook-down chain. With boats dragging in front of each other and behind, inevitably rigs got tangled, and occasionally there was a big mess, with cussing and arguing about who had the right of way. Both boats had to deck their nets and untangle their cables and go back to dragging until they ran out of space and ice. Then it was back to the fish house to unload, and then we got back out there and dragged some more.

When the bay shrimp were all caught and the rush was over, the *Lady Murle* headed into Raymond's Seafood to sell her catch. The trawler had to wait for four hours, tied up beside three other boats, before it was her turn to unload. Bucket after bucket of shrimp was hauled out of her hull and dumped on the screeching, smoking conveyor belt that carried them into the processing plant. There they were dumped out on large tables. Women called headers separated the heads from the bodies with flying fingers and fed the shrimp into a machine that graded them according to size.

When the entire catch was unloaded, the grimy, tired deckhands crawled out of the hold, looking irritably at Edward, who was enjoying his position as captain, walking around with a glass of bourbon and joking with the other skippers. After the catch was weighed, the captain and his deckhands waited in the office for Raymond to figure up their checks. Louie, a new deckhand making his first trip, earned $900, and Johnny earned $1,400. The captain grinned as he looked at his check for $11,000. It was enough to pay the shipyard bill he still owed from the year before, buy some new nets, and still have grocery money left.

With a drink in his hand, and celebrating, the skipper was feeling no pain. A good skipper was respected no matter how drunk he got. A man who had his own boat, led his own life, and worked for no one except himself was respected by everyone in the fishing community.

The next morning, the little boats were still scrapping, trying to round up the few shrimp that remained in the Mississippi Sound, as the *Lady Murle* went past them to join the other big trawlers in the open Gulf. Johnny had a $900 hangover, and Louie had a black eye, but they were happy. Edward headed southwest, hoping to find bigger shrimp. Far out past

Horn Island, away from the gear restriction zones, they set both nets to drag over "Sea Liver Hole," an old standby. Sea livers were actually "sea pansies," a small soft purple coral that glows in the dark. But there were no pansies or shrimp. All we found were large masses of stinging nettle jellyfish. The minute we opened the dripping net, the round purple blobs with their long trailing tentacles sloshed out on deck, and it rained fire. The mound of struggling fish that weren't already dead from being stung to death beat their tails as they swilled in a matrix of viscous liquid. They splattered our faces, necks, and arms with mucous and stinging cells.

"I'm burning up! Goddamn!" cried Louie in agony as red blotches appeared on his face, the mucous membranes in his nose became inflamed, his eyes swelled half shut, and he began sneezing and coughing. He tried to pick out the shrimp, but it was hopeless. Finally Edward ordered them to open the scuppers. When the boat leaned, the entire catch, shrimp and all, swilled messily into the sea. Nothing was a prettier sight than to see it go.

Suddenly I had an image of a cafeteria where smiling food servers dished out eight or twelve shrimp at a time. Shrimp laid out under lights, crispy brown and delicious, give no hint of what it takes to catch them. Shrimping is a grueling life of catfish spines, stings, bites, seasickness, and no sleep. You have to love it to keep doing it, and this young deckhand, fresh out of high school, was wondering if he loved it all that much. As we ran down the coast, Edward heard other boats complaining about the hot jelly; it seemed to be everywhere. But he also heard a rumor that a few boats were catching shrimp

50 miles to the west. It was a long way to run and a lot of fuel to burn.

The Florida shrimpers discussed the possibilities on the radio, trying to figure out what to do next. "It's too calm," Captain Joe Morgan Cruse declared. "What we need is a good blow to break up this jelly and kick the brownies out of the ground!"

Weather affects their behavior. When it's flat calm, the shrimp go into the ground and sleep. But when a storm comes, and there's a change in barometric pressure, they start moving. When there's a hard southwest blow, the water becomes turbulent, and the bottom gets stirred up, shrimp respond to wave surge and pop out of the ground. The rougher the water, the more they move around, whether brownies, hoppers, Key West pinks, or white shrimp. Perhaps the turmoil simply washes the slumbering shrimp out of their beds. Or it could be that they pop out in anticipation of the multitudes of sea worms, amphipods, and their other favorite prey items that are exposed as sediments are churned into solution.

The shrimpers knew what science is just beginning to learn: shrimp and other sea creatures respond to wave surge. When it's so rough that you can't stand up, when it's dangerous to be on deck, and accidents happen, that's when you catch them. When the swells are so big that they snatch the otter doors off the bottom. That's when the trawls can come up bursting with shrimp. The wave surge scours the bottom and churns the sediments into solution, and the worms are exposed. The shrimp pounce on them with frightening ferocity, picking and tearing apart their prey with their tiny but surprisingly powerful pinchers.

After the storm passes, the tender exposed worms and other creatures try desperately to rebury themselves, but it's often too late. Crabs grab them, fish eat them, and shrimp pounce and skewer the worms with their first two front legs that end in sharp points and have no claws. Fanning the mud with their beating swimmerets, they dig the recalcitrant worms out of the mud, grasping them with their tiny but powerful pinching claws, madly beating their swimmerets as they drag them up into the water column.

Another disembodied voice, belonging to Willie Smith on the *Miss Joysea*, broke in and said, "Be careful what you ask for, Captain Joe. You might get more of a blow than you want."

"That's why I named all my boats *Stormy Seas*," Captain Joe shot back. "It takes a good blow to make them hoppers and brownies run."

In a flat calm sea, with hardly a ripple, and pink and purple jellyfish pulsing along the surface, the *Lady Murle* left Horn Island and Cat Island in the distance and headed for Chandeleur, thirty hours away. The sun sank into the western sea, and Edward sat at the wheel, drinking coffee, looking at the glowing stars above. At two in the morning, Johnny was asleep, so the captain looked around, saw there was no traffic to worry about, and let greenhorn Louie take the wheel.

Feeling important and full of responsibility, Louie sat in the pilot seat, overlooking the vast Gulf. Sitting beside him, we noticed the ghostly forms of creatures swimming on the surface, and we cast the spotlight on them. The orange eyes of shrimp glowed back at us. Dozens of them were all around the boat, swimming at the surface. I'd seen a few shrimp in the water column over the years, but nothing like this. Louie rushed into the back cabin, shook the captain awake, and told him what he'd just seen.

Sleepily, Edward took the helm, but he wasn't excited. He broke back the throttle and slowed down so his strikers could set the small try net. Fifteen minutes later it came up with sea robins, small croakers, cusk eels, a remora, and not a single shrimp.

"I'm not surprised," he said, taking a sip of hot black coffee. "When you see shrimp on the water like that, which ain't often, it don't mean they'll be on the bottom. One year, me and a bunch of boys from Panacea ran over to Louisiana. We saw shrimp on top of the water just like this, only shrimp by the billions. What you're seein' here ain't nothing; they were so thick you could walk on them, and that's no lie. But we couldn't catch them. We set over, all excited, dragged through them, and didn't come up with nothing!"

Then he added, "Joe Morgan Cruse caught them, though. He was running the *Miss Joysea* and put the bridle half out and had the doors in the wheel wash, with the cork line on the surface, and pulled through them and loaded down! But he's a master shrimper. That's why he's got all them boats."

It was almost dawn when the lights of a dozen trawlers appeared in the distance, and as he approached them, Edward ordered his deckhands to set the nets overboard. "Sometimes that daylight tow pays off better than all the rest. The shrimp have a flurry before they bed down for the day."

A southwest wind was blowing, and the waves were starting to churn the green waters, a clear sign that the weather was changing. The bleary-eyed captain looked at the clouds, turned on the marine weather forecast, and learned that a tropical storm was crossing the Gulf, a big disorganized system with 45-mile-an-hour winds. It was expected to hit Pensacola, 150 miles to the east, but that was anybody's guess. It could go west and bring some terrible weather.

Edward studied the masses of gray clouds moving rapidly across an angry orange sky. A storm could sink us under the wrong conditions, but it could also make Edward and his crew rich. By evening waves were churning the Gulf of Mexico and whitecapping. The seas broke over the bow, and the boat pitched and rolled as the crew dropped the nets. Inside the cabin we waited with anticipation, hanging onto the benches and trying to eat in the rolling seas. The rains beat down with ferocity, and the wind moaned in the rigging, shaking the cabin. Soon it was time for the big nets to come up. Putting on their yellow slicker suits, Johnny and Louie stepped out into the wind-driven rains and began winching in the trawls. If they kept dragging, the weight of the swollen bags would close the doors, and the net would simply dredge through the mud. That would chaff the bottom, and the catch would swim away. They couldn't stop dragging and just sit there, or the cables would tangle and get caught in the propeller (see Figure 6.3).

Figure 6.3 Edward Keith adjusting nets on an outrigger—a dangerous job

Courtesy of Gulf Specimen Marine Laboratories, Inc.

The first net, gorged with a ton of marine life, became a swinging punching bag with catfish spines and stingray tails protruding. The decks were slippery, and the sea was pitching. Johnny fell flat on his back but wasn't hurt, and fortunately the deck wasn't full of catfish. On another boat, a deckhand got his foot caught in an outgoing lazy line and was dragged through a pile of catfish before he was snatched over. The captain managed to stop the boat, and they rescued him before the nets dragged him under, but the Coast Guard had to airlift him to the hospital. We all knew shrimpers who had broken legs and become crippled; others had been lost overboard. One lucky fisherman was knocked overboard and managed to grab on to a tow cable and pull himself up one stormy night.

Grimly Edward strode forward and tackled the swinging bag. With a snatch of the release rope, it disgorged shrimp all over the deck, nine boxes in one tow. We shrimped through the night and kept shrimping through most of the day. With skies overcast, the fickle brownies that shun the sun gave up their nocturnal behavior and stayed out, dancing around the bottom. After three days straight, Edward could no longer keep his eyes open, so Johnny took the wheel, while a soggy Louie fought his seasickness as we helped him cull off the stern. Two days later the sun came out and the seas calmed as the tropical storm went inland and once again the Gulf turned into a bathtub. The shrimp settled back down into the mud and went to sleep, and the catch dropped to two boxes of brownies a night.

The flurry was over and the fleet moved on, looking for shrimp farther west. Some of the fleet went offshore to drag in a hundred feet of water for jumbo shrimp. These were the great gravid mama shrimp that went fifteen to the pound. Through the translucent carapaces of the females, one could see the greenish-yellow eggs inside. The males carried white sperm packets on the outside of their abdomens. When the moon was full, and it was dusk, they would be ready to mate and shed their eggs. There weren't many, but the jumbos fetched the highest prices. Why people paid more for them wasn't clear, since they're tougher and chewier than their tasty juniors (see Figure 6.4).

Figure 6.4 Gravid white shrimp

Courtesy of David Barfield

Catching the mama shrimp was controversial. The bay shrimpers complained that they should be protected, since they were the reproductive stock that provided the new crop each year. But fisheries biologists were ambivalent, saying that nobody had ever proved that there was much relation between spawning and recruitment. Since a single female shrimp produced between a half million and a million eggs, some believed that a large number of spawning shrimp wasn't necessary or critical to repopulate the ecosystem. Other biologists disagreed, saying that the uncountable trillions of larvae were necessary to support the food chain.

The crew of the *Lady Murle* was happy to see the two one-hundred-pound boxes of six-dollar-a-pound jumbos, even if they were embedded in a ton of bycatch. The steady rumbling of the diesel throbbing through the stack mingled with the cries of the gulls that flocked over to eat the discards. Those shrimp boats were a spectacle with their avian clouds overhead, screaming and screeching until you couldn't hear yourself think. Every time a shrimp boat hauled back, it seemed that every brown pelican for a hundred miles beat its wings over to partake of the feast. They charged in like bombardiers, crashing into the water, and swam alongside like ducks on a pond, waiting to pounce the instant the crew pushed the fish through the scuppers (see Figure 6.5).

Figure 6.5 Seabirds feeding on bycatch

Courtesy of the State Archives of Florida

Seabirds were among the biggest beneficiaries, but porpoises and sharks also followed along, becoming more excited with each shovelful of bycatch that went back into the sea. Forever on the move, the sharks swam with grace and beauty, their never-tiring tails undulating. Following the boat, porpoises blew and huffed. Swarms of jacks, redfish, dolphins and even sea turtles followed the boat, eating the discards. Jack crevalles struck and splashed as the dead fish sank into the murky depths. The moment a stingray or fish hit the water, sharks rose from the depths to grab it. They were eating machines. Ten thousand or more diverse fish, crabs, and other wonders of the deep were shoveled through the scupper holes each time the catch was dumped on deck.

After several weeks at sea, forty boxes of shrimp were stashed in the ice hold. Everyone smelled fishy, and our clothes were thick with dried slime. Johnny showed Louie the trick of washing clothing by dragging it in the try net, the miniature trawl that could be quickly lowered and raised to see how many shrimp the big nets would probably catch. The deckhands killed time by fishing with a rod and reel, or they dropped a big shark hook over the side. Since sharks were always browsing around the shrimp boats, catching one was almost predictable. What wasn't predictable was the time a fourteen-foot tiger shark was hooked on a Texas boat and, after a massive struggle, was finally winched aboard. When the deckhands cut it open, the remains of a man fell out.

Sometimes the monotony of waiting at anchor was broken when a speedboat with sports fishermen aboard came by, wanting to trade chum for beer. In the old days, before shrimpers and sports fishermen became bitter enemies over bycatch issues, a sign language had developed between the deckhands and the sports boats: if a man came by and held out a six-pack, the deckhand waved him forward and exchanged a barrel of trash fish saved out of the catch specifically for bartering. The captain didn't mind and even encouraged it, until his shrimp were traded. But as red snapper stocks declined dramatically and the National Marine Fisheries Service imposed quotas on recreational fishermen, enmity grew. The recreationally oriented Gulf Coast Conservation Association went to war. In the name of saving the Kemp's ridley sea turtles nesting beaches, it succeeded in imposing a battery of regulations, ranging from limited entry to permanently closing the inshore waters of Texas, and the friendly bartering ended.

Each night the crew of the *Lady Murle* culled mounds of fish, picked out baskets of flounder, whiting and butterfish and layered them with ice along with the shrimp. With July temperatures near 100 degrees, the ice in the hold was melting fast. The shrimp caught at the beginning of the trip were starting to go bad. Unless the perishable shrimp were "broken," the digestive enzymes in the heads would rot the tails. Edward joined us down in the hold, heading shrimp for a backbreaking nine hours. The beheaded shrimp were washed thoroughly with the deck hose to remove any slime, and then they were stashed in their own compartment down in the hold. They were kept separate from the recently caught heads-ons.

The *Lady Murle* had been at sea for sixteen days and inevitably, land fever developed. The young men talked about the good times they had last year in Biloxi at the bars. "No way," Edward told his crew. "We're not going in."

"Edward's a better shrimper than Bobby, Glenn, and the whole bunch put together 'cause he works hard; he's steady go," Johnny explained to a despairing Louie. "He pushes himself, his crew, and the boat, and he's gonna make it. He'll go round the clock, beat his way out in rough weather. But he plays hell with his deckhands; he expects them to work as hard as he does."

But when Captain Edward heard on the radio that the price of shrimp had dropped and there might be yet another cut before long, he changed his mind. "All right, clean her up. We're going in!"

With a joyous shout his crew began cleaning up the boat, washing down the decks, hauling in the rigging, tying down the otter trawl doors. We scrubbed the cabin, washed the pots and pans, cleaned the grease off the stove, and got the boat shipshape. The nets were hung out to dry while the boat steamed ahead, over the open sea. A day later the green marshlands, trees, and buildings of Pascagoula came into view. Finally we were in the industrialized river, heading up the channel, past the grain docks to Raymond's fish house to unload. Edward and crew were ready to take a break and have some fun before heading back out again. I drove back to Panacea with a load of specimens for our aquarium.

Edward was running a smaller boat one frigid winter night in 1996 off Alligator Point when a single huge wave swept over the stern, flooded the engine room, and sank the boat. While his two deckhands survived, Edward died of hypothermia, and it was three months before what was left of his remains washed ashore. Today a monument sits in his memory in Panacea, Florida, next to the sea he loved. "If the shrimp eat me," he often told us over a bottle of whiskey at our house, "that's only fair. I've caught and eaten enough of them. It's finest kind!"

7

Rock Shrimp and Spotted Prawns

Anything humans can't use, or understand, they label as trash. For years scientists called the parts of the genome that appeared to do nothing "junk DNA" before they discovered how important it is in controlling gene expression. Medical people declared the appendix a vestigial or useless organ and cut it out, although recent research has demonstrated that it serves as a reservoir for bacteria essential to digestion. Shrimpers call all the marine species in their nets that aren't worth money "trash."

For years little colorful hard-shelled rock shrimp were pushed through the scuppers along with the rest of the bycatch. There was no market for them, even though the shrimpers often took them home to "cook up a mess to eat."

Shortly after I (Jack) returned from Madagascar and started a marine specimen collecting service in Panacea, I was back on shrimp trawlers in Apalachicola, gleaning through their catches and picking out octopuses, sea horses, and a plethora of other sea creatures.

When white shrimp season ended in November, most fishermen left their boats docked for the winter, waiting for the hoppers to run in April. Dave Silva was one of the few captains who went looking for hoppers in deep water during the winter, and he wasn't afraid to risk his nets dragging in rough sponge bottoms. I was aboard one trip, when he hauled his nets from 70 feet of water and buried the deck in a beautiful little hard-shelled shrimp I had never seen before. He called them "rock shrimp" because of their rock-hard shells. They were gorgeous creatures, purplish brown in color, with white stripes on their legs.

"Take some of the rock shrimp home and boil them up," Dave said. "They're real good. We can't sell them; there's no market for them." Neither was there a market for the strange-looking bulldozer lobsters that flipped and crawled slowly around on the deck like their namesake. The rock shrimp were fantastic, with an unbelievable taste that was more like New England lobster than shrimp. And the bulldozer, or slipper lobster, was just as good as, if not better than, the spiny lobsters that were caught in the Keys. Dollar signs danced in my eyes (see Figure 7.1).

Figure 7.1 Rock shrimp

Courtesy of the Florida Department of Agriculture

We started marketing rock shrimp to fancy seafood restaurants, fish markets, and grocery stores. I printed brochures and hired a friend to peddle them door to door in high-end neighborhoods. Gourmets loved the taste but complained that the shells wore out their fingers trying to peel them. Furthermore, the gut was filled with sand and grit, since rock shrimp love to eat tiny mollusks and worms that live on the coarse shelly bottoms. So sales were slow at first.

Hoping to find a big sale, I took samples to the old Fulton Fish Market in New York. What a place! Before the new sanitized one was built, there were stalls of every imaginable seafood. At four o'clock on a cold winter morning, rough-looking men wearing boots dragged titanic flounders and eight-foot-long headless swordfish across slimy snow-covered sidewalks with boat hooks. Amid the noise and bustle, peddlers hawked their product in a feeding frenzy of restaurant buyers bidding and arguing over the price. There was no end to the seafood that was brought in. Trucks were lined up everywhere, their engines rumbling as teamsters unloaded them. Men warmed their hands over a fire that burned in a metal drum beneath the subway overpass and then unloaded hampers of bubbling live blue crabs from Virginia. Some stalls specialized in mussels, oysters, clams, and cans of shucked scallops. One truck from Maine unloaded hampers of green sea urchins with their spines still moving, and bags of live lobsters. Another bore big black cod from Canada—before the fishery collapsed. There were striped bass from Long Island; whiting and grey and speckled trout from the Carolinas; grouper, snapper, and sharks from the Gulf of Mexico. Big red scorpion fish lay gleaming on beds of ice, their eyes glazed under the lights, all fascinating to my eye. I went from stall to stall showing my rock shrimp. No one wanted them. "Look, man, I sell fish and shrimp. I'm not a missionary. I can't sell these," one vendor said sympathetically but firmly.

Then I came to a stall where buyers were lined up filling bags of big white shrimp from Florida. To my surprise, there was an unsold bin of headless rock shrimp. The vendor shrugged, "No one wants them; no one knows what they are. I haven't sold any!" They came from Rodney Thompson, who built the first fiberglass 73-foot shrimp trawler in the Western Hemisphere and then almost went broke trying to shrimp it. "One lonely afternoon, we tied up at Port Canaveral next to the NOAA Research Vessel *Oregon II*," he later wrote. "Captain Barret looked at our empty nets and grinned, 'Do you want to make a million dollars? I'll show you how!'

"We were hesitant, but we were starving. The next day, in the wake of the *Oregon II*, we found ourselves 20 miles east of Melbourne, Florida. We dropped our nets and after an hour decked over one thousand pounds of 'peanuts' or 'hard heads.' The captain of the research vessel (who also helped develop the royal red shrimp fishery) said, 'If you can figure out how to sell those 'peanuts,' you'll be a millionaire.'"

After heading the rock shrimp, Thompson shipped them to fish markets throughout the country. He ran into the same problems trying to sell them until he designed a machine with a high speed spinning blade that split the shells open. The shrimp could then be broiled like lobsters. He sold his boat-building business and started the Dixie Crossroads Seafood Restaurant, in Titusville, Florida, specializing in rock shrimp, and he indeed made his millions.

Broiled in their shells like a lobster, they quickly became a popular seafood delicacy. We, on the other hand, nearly went broke. While we were pioneering the market on Florida's west coast, the fish house owners, who controlled the fleet, watched and waited until the market took off. Then all our customers were suddenly buying direct at a lower price. We were stuck with three thousand pounds of rock shrimp in the freezer. A dealer took them off our hands with a check that bounced. The day we finally broke even, we had a great celebration and vowed to get out of the seafood business once and for all.

Various seafood processors tried to get into the market by using the Laitram shrimp-peeling machines. They peeled and deveined white, pink, and brown shrimp using a series of oscillating sandpaper-covered rollers that scraped, pinched, and jiggled the shells off the meat and washed away the vein. This ingenious machine was invented by a sixteen-year-old boy, J. M. Lapeyre, in Louisiana in 1943 when he stepped on a well-aged shrimp at his father's shrimp processing plant and it squirted cleanly out of its shell. Inspiration struck, and he ran some shrimp through his mother's wringer washing machine. At first, the slime gummed up the rollers, but when he added water, he started popping out shelled shrimp. What his mother thought of the experiment was not recorded, but the family soon founded Laitram Machinery. It revolutionized the shrimp processing industry, with shelling machines producing a thousand pounds an hour and doing the work of thirty to a hundred fifty people peeling shrimp by hand.

But the machines fared badly with the tough shells of rock shrimp, so Rodney Thompson kept his rock shrimp market. For the next ten years his fleet of rock shrimp boats produced about ten million pounds of processed shrimp a year, employing dozens of women who worked his splitting machines and cleaned them by hand.

In the 1980s, the Pascagoula Ice Company successfully modified its Laitram machines, mechanically mass-processed the rock shrimp, and turned the crusty hard-shelled rock shrimp into tasty bites of lobster meat. The gold rush was on as other processing companies followed suit and the rock shrimp beds in the eastern Gulf and the east coast of Florida were under full attack from boats from Mississippi. Dozens of trawlers arrived on Florida's east coast, some of them 110 feet long, dragging as many as four 60-foot flat nets at one time. They could catch and hold more than five thousand pounds of rock shrimp a day in their freezers. Catches peaked in 1991, when 40 million pounds of rock shrimp crossed the docks in Florida, and then catches rapidly declined.

Along with royal reds, these deeper cold-water shrimp reproduce more slowly than warm-water shallow pink, brown, and white shrimp. The little hard-shell shrimp live almost twice as long as their softer-shelled cousins, produce fewer eggs, and grow slower, making them more vulnerable to overfishing. Desperate to maintain their cash flow and make their huge boat mortgage payments, the super-trawlers moved ever deeper into the rock shrimp spawning grounds further south. They caused enormous damage to fragile deep-water *Oculina* coral reefs that grow in bizarre spires called “cones” or “steeples” because they look like church steeples. The east coast of Florida is the only place in the world where these unique structures grow, providing a haven for marine life and structure for spawning rock shrimp. Using modern bottom plotters and GPS, the trawlers could safely crisscross the sides of the steeples without hanging up their nets. But the heavy chains and doors devastated the seafloor, much like the trawlers did off Apalachicola when they stripped the sponge bottoms. Inevitably, a few years later the fishery collapsed. There weren’t enough shrimp left to pay the costs of a fishing trip.

Today the Oculina reefs are off limits; the number of boats permitted to fish is strictly controlled and landings have stabilized. Since marine bottoms are in the public domain and marine life is a public resource, government regulators control fishery harvests. They now require shrimpers and other fishermen to use an electronic vessel monitoring system that puts out a signal when they're approaching the Oculina banks or other closed areas. The signal is uploaded to a satellite, which is tracked by NOAA Fisheries technicians. It warns the boats that they are approaching a closed area. If the fishing boat continues, the Coast Guard is notified and dispatched.

Across a continent from the Florida rock shrimp grounds, we stepped aboard Bruce Gillespie's 26-foot-long *Commander* sports vessel to head out on a prawning expedition from Courtney, British Columbia. Unlike the boom-and-bust rock shrimp experience, a more sustainable fishery developed in the Pacific Northwest for seven species of the cold-water shrimp and prawns in the family Pandalidae.

After days of gloomy, cold wet rain, the rising sun was indeed a warm sight. For a few moments, the inland sea was pearly white with the early morning light. Seagulls flew across the skies or perched on the water, waiting for fish to come up. We watched the sun rising over the mountains, bathing the shoreline in light, bringing out the fall colors of festive and bright yellow leaves, and chartreuse green. The scenery dominated in this place of commercial and sports fishing, where aquaculturists grew clams and oysters in rafts, sea lions climbed on them to sun in the afternoon, and First Nation (Native American) fishermen set their gill nets to catch chum salmon. We were after spot prawn (*Pandalus platyceros*), which is the largest of the pandalid, or northern pink shrimp. Its geographic range extends from Southern California up to Alaska's Aleutian Islands and around to the Sea of Japan. The spot prawn fishery began in California back in the 1930s, when Monterey fishermen started landing spot prawns from their octopus traps. In the 1970s, it spread up and down the Pacific Coast from California to Alaska.

"We're a month early," said Bruce. "Most people around here don't start prawning until November or December. They want those Christmas prawns to put in their freezer, and that's when they start. The salmon are running now, or the guys are in the woods deer hunting. Later on, when the prawns are bigger, and running, there will be forty or fifty floats out here, and you have to watch out that they don't get tangled with yours. Then the commercial season starts in February." Commercial shrimpers can fish two hundred fifty traps and can fish from sunset to sunrise during the two-month season that begins in February. They load down with prawn in a short, vigorous, and highly profitable season. About five million pounds are landed each year. The recreational fishery is subject to daily catch limits, but the commercial fishermen have none.

It was a short thirty-minute ride over flat calm water before we arrived at the prawning grounds. Over the years while working on scientific collecting expeditions or on writing assignments, we had shrimped in South America, Africa, Malaysia, and China with flat coastal landscapes and marshy muddy rivers that turned into a faceless horizon of rolling seas and whitecaps. Here it was calm, deep, and protected. Somewhere, fifty miles or so away, across mountains and forested landscape, this sound connected to the Pacific Ocean. But as we set out in Bruce's little boat, it was hard to think of it as anything but a mountain lake. Long ago, a glacier gouged out the land, and the sea flooded in. As Bruce unstowed his prawn traps to get ready for the day's fishing, the sea, if it could be called a sea, was calm, with barely a ripple on the surface. But it reached to depths of 200 feet or more of freezing water—too deep to span with bridges. So ferries ran back and forth to the opposite shore.

Everything on this charter recreational boat was organized and neat, from the dozens of diverse fishing lures in a rack on the cabin walls to the fishing rods that were held in place inside the gunnels. Most of Bruce and Nancy's clientele were after salmon, trout, and other large fish, not prawns. The few who saw his ad on the Internet and went prawning did so more to see the scenery and be on the water.

We carried aquariums, cameras, tape recorders, notebooks, and Greg Jensen's book, *Pacific Coast Crabs and Shrimps*. A professor at the University of Washington, Jensen spent a lifetime studying the crustaceans of the Pacific Northwest. "One doesn't shrimp in Canada," Greg corrected us. "One goes prawning. In most places the names are interchangeable." Here in the Pacific Northwest they prawned with traps, and at the height of the season the sea was polka-dotted with their floats.

Most of the prawn industry centered on the smaller *Pandalus eos*, which are caught in great numbers by commercial trawlers dragging nets over the bottom. Pacific prawns live in the cold depths, where the water temperature hovers around 40 degrees year-round. If you fell overboard, you wouldn't stay alive much more than fifteen minutes, but the shrimp and a whole cadre of fish and Dungeness crabs like it down there. They differ from southern pink, brown, and white shrimp in that they have a large carapace and a relatively small tail, but because they are shrimp, they taste good.

Most recreational fishermen simply broadcast their traps and hope for the best, but after fourteen years of fishing the same waters, Bruce knew the bottom contours, submarine gravel beds, and rocks intimately. He carried eight plastic-coated wire traps that he had designed and made himself. As recreational fishermen, we were allowed two per person. Some of his traps were so close to shore that, as the tide dropped, exposing flats along the shore, we could see the rocks and brown clumps of seaweed hanging limply. If this were the shallow bays of Florida, we'd be running around, but here we were looking for a different shrimp in a different world.

Bruce unpacked the frozen plastic drink bottles filled with bait. They were riddled with holes so that the juices could flow out. His bait was secret, but he admitted to using a blend of commercially available aquaculture "starting" pellets. Some of it was fish meal, mixed with brown fishy-smelling salmon chow from the salmon farms. "The shrimp love it," he said.

Nancy ran the boat while Bruce studied the fathometer, looking for just the right spot to place his traps. Down in the cabin, he showed us the depth recorder. "You see how it drops off on the south end of the line," he said. "The bottoms are different, and so are the shrimp, in just a short distance. If you put traps at random, you catch a few, but not in volume. I know from the sound what's on the bottom; you don't see it on the charts." He checked the contours and the depth of the water and watched the shore for landmarks.

Spot prawns tend to inhabit rocky or hard bottoms, glass-sponge reefs, coral beds, and the edges of marine canyons. Perhaps they find food there, or larval transport carries them to a particular spot, but there they stay. Their distribution is notoriously patchy, and no one knows why.

As we approached the spot where he planned to set the traps, Bruce hopped back and forth, up and down the steps leading into the small cabin, studying the fathometer, looking for the right spot that lay below the transparent blue water, out of sight.

"Neutral," he called out, and Nancy pulled the gear shift lever back so that he could lower the trap into the water. Typical of all the shrimp cages used along the Pacific Northwest, all the trap openings pointed upward. "We make it easy for them. The shrimp crawl all around the outside of the trap, trying to get to the bait, then go up the shoot and fall off into the cage." People had tried these traps in the Gulf of Mexico and South Atlantic for catching penaeid shrimp, but they never really worked out.

Down into the blue water went the two baited wire cages, the rope uncoiling from the milk carton. Finally the big red inflatable float was tossed overboard with a splash. Fifty feet from where the rope was tied onto the trap was a four-pound weight, which acted as a shock absorber. "Wind will make the trap move," Bruce said, "when it pushes against these buoys. The shrimp won't go in, but this lead weight keeps the traps from moving on the bottom." The inflated red buoy bobbed on the blue sea behind us as he headed for the next set. The first two traps settled to the exact spot where he wanted them to be, next to the rocks, where, according to the graph on the fish finder, the bottom rose. After the last of the traps were deployed, our guide cut the motor and we drifted, waiting for the hour and a half until it was time for the traps to be pulled. There was no leaving the traps in the water overnight.

"We can only leave them soak for an hour or two," Bruce explained, "because predators, like giant sun stars, octopuses, and other fish will get in and eat the shrimp."

Deep-sea rattails, six-gilled sharks, rock fish, and giant cold-water octopuses lived in the glacially carved submarine terrain. Salmon, sea lions, and a plethora of other marine life made their way in from the fifty-mile-distant sea, or grew up in the deep waters. When schools of herring came in to spawn, the water became cloudy with milt. Steelhead salmon ate the herring and gulped down schools of anchovies in a fish-eat-fish world. While we were waiting for the traps to come up, Nancy brought out chunks of smoked salmon, which they had caught and prepared themselves. "We catch five or six salmon a year, and that's enough to live on," she said, "but a lot of these sportsmen are greedy and catch their limit of two a day. The fish are huge; there's no sense in doing that."

Killing time, we set off exploring, leaving our big red buoys sitting quietly on the surface of this millpond of a sea. We passed rafts of floating logs with more than a hundred barking and croaking sea lions lounging about in the sunshine that was so rare on Vancouver Island's cloudy rainy coast. They complained loudly as we encroached on their comfort zone, many flippering and heaving their way into the water with a splash and swimming around with their heads up. Some were on their backs with their flippers in the air.

"They're so abundant they're a nuisance," Nancy said. "They eat tons and tons of fish. They go upriver and gorge themselves on coho salmon. Anything they can get their jaws around, they eat."

When it was time to pull the traps, Bruce caught the buoy with the boat hook and ran the line through his homemade puller. It was smaller than commercial models, designed to fit his boat and give lots of room. He was embarrassed when the trap came up empty. "I must have had this one off the edge. We're fishing on a small patch of gravel bottoms," he explained, "and if we get off, we don't catch them. The traps are a hundred feet apart; we'll do better on the next one."

He was right. Minutes later, the next trap, attached to that same line on the buoy, emerged from the 200-foot depth full of shrimp. Inside the wire, we could see pinkish-orange jumping prawns, thrashing and kicking inside the trap. There was also a big yellow multilegged creature, a sun star, hanging on the surface. "We don't want him," Bruce said. "They're nasty. The sun stars come in here to eat the shrimp. Sometimes we find them inside the trap, and there'll be nothing but shrimp skeletons if you leave them too long."

He opened the trap lid and shook five or six pounds of jumping and flopping red shrimp onto the culling table. We counted seventy-five spot prawns and one coon-striped shrimp. The shrimp were a dazzle of red and pink, with delicate candy-cane-striped legs. Their bodies had an almost luminescent translucence. Looking at the lack of bycatch, we had to agree with the Monterey Bay Aquarium's assessment that the northern pink shrimp trap fishery is one of the least damaging and most sustainable.

We had handled hundreds of shrimp of many species. We knew the spikes of pink shrimp, browns, and whites, the sea bobs, and the broken-backs. Out in the Gulf of Mexico, nearly two hundred miles from shore, we had caught deep-sea royal red shrimp off the decks of a 90-foot steel hull, working in blue water. We handled rock shrimp with hard shells, and big fresh-water prawns from the rivers in Costa Rica. But nothing was as evil-looking as the rostrums or head spines of these cold-water pandalid shrimp that projected from their heads. Needle-sharp swords, they were wickedly curved, like the knives of Gurkha soldiers. Looking at them we understood why early biologists named this spine after the battering rams mounted on the prow of ancient Roman war ships. Of the seven species of *Pandalus* in the area, species that have shorter rostrums generally live under cover, beneath rocks or logs, while those with long rostrums live out in open waters. According to Greg Jensen, the rostrum makes them hard to swallow, and they also use it defensively. When a fish grabs a shrimp, the shrimp goes limp while the fish tries to swallow it headfirst. But with that spiky rostrum and antennal scales, the fish releases the shrimp for a second, trying to get another bite, whereupon the shrimp has a chance to jump away and escape (see Figure 7.2).

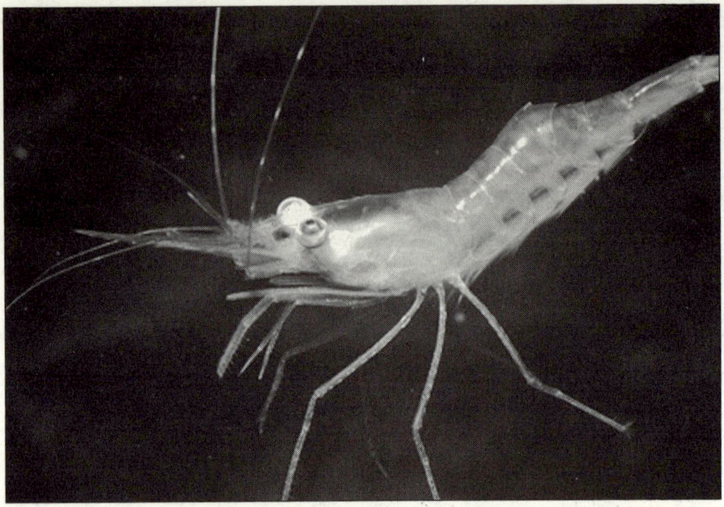

Figure 7.2 Northern pink shrimp

Courtesy of Greg Jensen

Looking at these daggers and remembering stories of people who were stuck by them and got infected, we gingerly placed the shrimp in the glass tank, trying not to get stuck. It was hard to tell the species, as they all looked alike. Only a specialist could tell the difference. The prawn's legs were like candy cane stripes, while the coon-striped shrimp had white spots on their abdomens. A few soft-shell shrimp were in the catch, which meant we wouldn't have to shell them before eating them. When a prawn or shrimp gets ready to molt, it grabs onto a rock, and in thirty seconds it rolls out of its shell. The empty shell looks like cellophane.

Like many other caridean shrimp species, these had a confusing sex life. The spot prawn is called a protandric hermaphrodite, which means that it starts as a male and then becomes a female. It generally reaches sexual maturity as a male by its third year. After it mates as a male, it enters a transition phase and in the fourth year changes into a female and then mates as a female. When the fishermen go after the larger shrimp, they catch out the reproductive females (see Figure 7.3).

Figure 7.3 Spot prawn

Courtesy of Greg Jensen

Nancy was on deck, scrutinizing the catch for the big female prawns to throw back. She looked for the pinkish-orange mass of eggs beneath their carapace. "The eggs are just starting to show, and they haven't really fluffed out," she said. She tossed them back as Bruce moved the boat back over to the gravel patch of bottom 200 feet below to pick up the next float. Watching them tail-flip their way back down into the sea, we wondered about their chances of getting to the bottom alive. "I don't know. I think most of them make it," Nancy said. "Not many fish feed on them up here on the surface. You don't ever find them in salmon, but the rock fish on the bottom love them."

"Sometimes we get big octopuses in the traps," she said. "I don't know how they squeeze into the openings, but they do. Then you really see the shrimp shells pile up. They eat them as fast as they can catch them."

Aboard the little sports boat, Bruce broke the heads off the shrimp that didn't have eggs. We had them for lunch. The meat was a firm dazzling white—whiter than the whitest shrimp I ever saw in the South Atlantic. But finding them in the market is rare, since the biggest prawns are quick-frozen at sea and exported to Asia. Preparing them properly and aesthetically arranging them in the packaging is an art form. Down the bay, five or six First Nation gill netters were catching salmon. Although Native art depicts salmon, killer whales, and birds along the Pacific Northwest, there's not a single totem pole of a shrimp, or a shrimp effigy. Just as on the Gulf Coast and South Atlantic, shrimping is a recent fishery.

As the afternoon sun was starting to set, we headed back to the dock, over the clear glassy deep water. Down below were shrimp—shrimp of another color, shrimp of another species—another shrimp in another world.

8

Turtles, TEDs, and Troubles

With a jerk of the tail bag rope, a deluge of fishy, bulging-eyed water creatures tumbled out on the deck, a huge loggerhead sea turtle on top of the pile. Dock lights gleamed off her scaly yellow flippers and brown barnacle-covered shell. Gasping for breath, she slowly blinked her heavy-lidded eyes. As she gained strength, she began plowing through the squirming catch of croakers, crabs, and shrimp like a bulldozer.

The *RV Georgia Bulldog*, a 73-foot shrimp trawler run by the University of Georgia, was dragging its nets along the Georgia Sea Islands at the height of the sea turtle nesting season. One of the trawls had a turtle excluder device (TED), or "turtle shooter," as the shrimpers called it. The other was a standard trawl without a grid of any sort to block the entrance. Sure enough, it caught a turtle.

Biologists from the University of Georgia grabbed the 200-pound female by her front and rear flippers and flipped her over on her back so we could tag her. She went wild, scattering the shrimp and fish with each pound of her flippers, snapping her powerful jaws that could crush a shell or a hand. After affixing a metal tag in her fore-flippers, the biologists dragged the loggerhead across the deck, hoisted her onto the rail, and heaved her over the side. She landed with a splash and rocketed off.

The great ponderous sea turtle has become a symbol of the world's environmental conflict, of uncaring pollution of the ocean and humans' disregard for other forms of life. The turtle offers itself helplessly as it struggles out of the sea to lay its eggs on land. Awkward and ungainly on land, it is a thing of beauty and grace as it swims in the sea. They drown in our fishing nets, are sucked up by dredges, and are ground into turtle burger, chopped up by boat propellers and blown to bits when obsolete oil rigs are removed by detonation. People eat them and their eggs in much of the world. They wash up dead on the beaches, often with their guts crammed with plastic bags, which the turtles mistake for their favorite meal, jellyfish.

Today all five species of American sea turtles are protected under the U.S. Endangered Species Act, but before 1966, sea-turtle meat was freely sold at restaurants. Shrimpers routinely butchered greens, Kemp's ridleys, and loggerheads. Deckhands made good pocket change selling turtle meat to the fish houses. Key West's turtle kraals were a famous attraction, where tourists snapped pictures as sailing sloops from

Nicaragua and the Cayman Islands unloaded hundreds of big green turtles at a time. But by the 1980s, times had changed. Conservationists were fighting to save sea turtles from extinction, and shrimpers were fighting federal proposals to require TEDs in their nets, which they feared would dump shrimp as well as turtles out of their nets.

There was little question that sea turtles were in trouble from shrimp trawling. Whenever the fleets of shrimp boats dragged close to the turtle nesting beaches of the southeastern United States, dead sea turtles soon washed up on the beaches.

Sea turtles can hold their breath underwater for hours, or even days in winter when they're sleeping on the sea floor, but they must come to the surface to breathe. However, when dragged in a trawl, or hooked on a swordfish boat's longline, they panic and soon drown. When the shrimp fleet left, the mortality stopped, and few strandings occurred. It was illegal to kill turtles, but the law was rarely enforced, and TEDs were not yet mandatory, so the responsibility lay with the shrimper. Often, when they caught a turtle, the crew hurriedly butchered it, stashed the meat in the refrigerator, and threw the shell overboard to hide the evidence. The next day they had turtle soup. Commercial fishermen who grew up eating turtle along Florida's panhandle coast considered turtle meat to be as much a part of their diet as any other seafood. The critically endangered Kemp's ridleys that nested on the beaches of Rancho Nuevo in Mexico had declined from 40,000 nesting females a day in the late 1940s to 6,000 in 1960 to 500 in the 1980s. Mexico, in cooperation with the U.S. Fish and Wildlife Service, had initiated intensive conservation efforts, stopped the harvesting of eggs, and stepped up nest protection. Hundreds of thousands of hatchlings made it to the water, but the mother turtles continued their decline. Death in shrimp nets was the chief suspect.

The U.S. Fish and Wildlife Service was responsible for nesting turtles on the beach, but the National Marine Fisheries Service (NMFS) took charge when the turtles were back at sea. The NMFS was charged with the conflicting responsibilities of protecting sea turtles and helping the fishing industry. NMFS scientists had spent years trying to help the people who shrimped, but now they were caught in the middle, with environmentalists demanding that they enforce the Endangered Species Act and crack down on the shrimpers. The NMFS began looking for a technological solution by creating a net modification that would allow turtles to escape while keeping the shrimp. Beginning in 1978, the agency spent over $2 million developing its TED, a panel that shunts big things (turtles, horseshoe crabs, large rays) out of the hole while letting shrimp pass into the net.

According to NMFS statistics, it was estimated that 48,000 sea turtles were caught each year on trawlers, and approximately 11,000 died. Of those, 10,000 were loggerheads and 750 were Kemp's ridleys, the world's most endangered sea turtle. The catch data came from 38,000 observer hours on shrimp boats, where 498 turtles were caught and 168, or 33 percent, were dead on deck. Some of those fatalities came from trawlers making excessively and unusually long drags—one boat up to ten hours.

We have published research on sea turtle biology, written books on them, and worked with shrimpers for over thirty years, and those numbers didn't match our experiences. We had personally witnessed dozens of turtles get caught, but only one was dead on deck. The rest, if the shrimpers didn't eat them (which they occasionally did), were thrown back alive.

While on an assignment for the *Smithsonian Magazine*, we asked the NMFS to let us interview the observers who witnessed such mortality, but we were refused. The issue was so controversial, we were told, that if their names got out, their lives would be in jeopardy. This seemed a bit strange, since these observers were at sea for weeks with the same potentially murderous shrimpers, counting dead turtles, and they never came to any harm. By sheer accident we met an observer in Port Arthur, Texas, who said that in two years of riding shrimp boats, he had not seen a single turtle caught. When asked about this, the NMFS said that wasn't unusual. Depending on where it is fishing, a vessel may work for weeks, months, or years without a turtle fatality, and some boats never caught them at all. Yet the cumulative impact of the fleet was considered serious.

At the time, there were 7,000 large trawlers in the South Atlantic and Gulf of Mexico and another 10,200 smaller boats. Even though smaller vessels made shorter tows, which give the turtles a better chance of surviving, they too drowned turtles. Mortality was especially high off the Georgia and South Carolina Sea Islands during white shrimp season.

Most of the shrimpers in the western Gulf of Mexico, where the bulk of the shrimping industry was found, insisted they almost never caught turtles in their trawls. The biologists argued that it didn't take a lot of mortality to impact the turtle population, especially the endangered Kemp's ridley. Unlike penaeid shrimp, which broadcast millions of eggs and live only a year or two, sea turtles were slow to reproduce and laid only a hundred eggs every two years after they matured, which could take twenty or thirty years. According to NMFS data, with shrimpers collectively making 300,000 trips a year, it would take only 1 percent of the fleet catching a turtle to have a severe impact.

By 1981, the federal fishing agency was promoting voluntary use of the device. In 1983 it began passing out free TEDs to encourage shrimpers to try them. Unfortunately, the first TEDs were heavy rigid metal devices. The fishermen quickly judged them to be dangerous, swinging violently overhead when nets came up in rough seas. They were also expensive, costing some $400 per net in an industry already stretched to the economic breaking point by rising fuel costs, overfishing, and competition from cheap imported Third World shrimp.

Arguably dangerous, expensive, and, worst of all, imposed from outside the closed, clannish world of the docks, the free TEDs were left to rust on shore. All the while the dead turtles continued to wash ashore, and environmental groups were demanding an end to the killing.

Later versions were lighter, safer, and cheaper, but nobody would touch them. Despite limited test results to the contrary, shrimpers feared that, under real-world working conditions, the lighter and cheaper TEDs would still cause them to lose up to 20 percent of the shrimp caught. By 1986, when the program for voluntary use of TEDs failed, the patience of the environmental groups had run out. The U.S. Fish and Wildlife Service and the Gulf of Mexico Fishery Management Council demanded mandatory use of TEDs. The NMFS then announced proposed regulations, but nobody on either side was satisfied. Notice of an impending lawsuit was then filed by the Center for Marine Conservation under the Endangered Species Act against the NMFS. Such a suit would almost certainly succeed and would probably force closures of the fishery (see Figure 8.1).

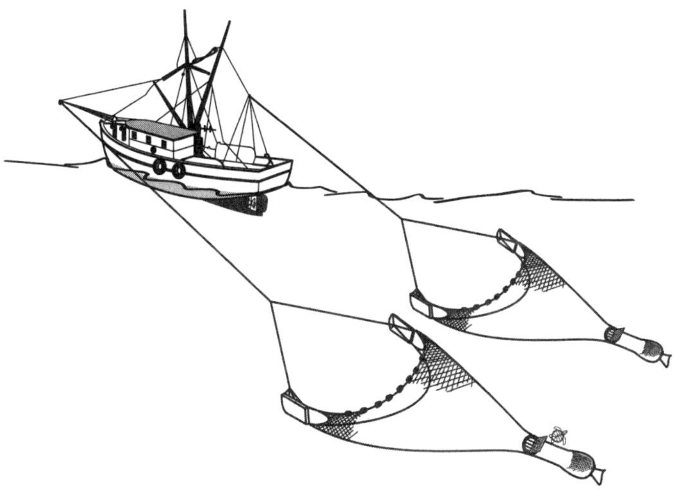

Figure 8.1 Turtle Excluder Device

Courtesy of Ben Higgins

Their attention finally engaged, industry representatives asked for mediation. The agency then offered a novel approach: all affected parties should sit down and negotiate their own settlement, writing the regulation themselves. A professional labor mediator was used to chair the sessions, and a series of four meetings were held. Environmentalists wanted immediate use of TEDs everywhere. Industry offered to limit the amount of time the nets would stay down so that turtles could be brought up and released before they drowned. While theoretically possible, this approach was unenforceable and therefore unacceptable. The heated arguments continued (see Figure 8.2).

Figure 8.2 A shrimper protests TEDs

Courtesy of the State of Florida Archives

When public hearings were held to require TEDs, the shrimpers rose up in rebellion. Thousands showed up in public auditoriums in towns like Biloxi to vow civil disobedience against what they perceived as a threat to their survival. Governor Edwin Edwards of Louisiana testified, "Perhaps some species were just meant to disappear," to overwhelming applause. He continued, "If it comes to a question of whether the shrimpers or the turtles are going to be in trouble, bye-bye, turtles." However, it was bye-bye Governor Edwards when he was later convicted on seventeen of twenty-six counts of racketeering, extortion, money laundering, and mail fraud.

Shrimpers are good at catching shrimp and lousy at public relations. After blockading major-port shipping channels in Tampa, Houston, and Port Aransas and hanging turtle dolls in effigy in resistance to using TEDs, they caught the public's eye, but the attention wasn't friendly. While all sides appeared equal at the table, in reality the environmental groups had the overwhelming firepower of the Endangered Species Act on their side. Industry representatives had only their power of persuasion and the knowledge that they might well be closed down if an agreement wasn't reached. Finally a rule was agreed upon to phase in the use of TEDs over a three-year period, 1987 to 1990.

When the Endangered Species Act came up for reauthorization, the war moved to Congress, which debated an amendment to give the industry a two-year moratorium on TED use. Again, the argument was long and heated. In the middle of it, the TED regulations took effect on January 1, 1988. The gear was required for about a month, until a judge ruled against the State of Louisiana's lawsuit to stop TEDs but granted an injunction on enforcement pending the appeal. The NMFS declared that only in Cape Canaveral, where loggerhead sea turtles swarmed into the shipping channel in incredible numbers, would the protection devices be mandated.

We went to Port Canaveral to go out with Paul Raymond, a fisheries officer, onboard a Coast Guard patrol boat to see how the laws were being enforced. Sea turtles had been killed by the U.S. Army Corps of Engineers for years while they were dredging the Port Canaveral Ship Channel. For some unknown reason, hundreds of loggerheads flocked there during the winter, buried themselves in the flocculent mud and went to sleep, and some were sucked into the draghead. It had been Paul's messy job to walk around the grates on the giant, smelly hopper dredge and look for turtle parts mixed in the fountain of black mud that reeked with hydrogen sulfide. If he or any of the other graduate students working on the project found too many turtle parts floating in the muddy milk shake, the million-dollar-a-day operation would be shut down.

To keep that from happening, the U.S. Corps of Engineers hired a shrimp trawler to catch turtles, tag and measure them, and move them miles away from the dredging operation. We went aboard the shrimp boat rescue vessel and watched graduate students and technicians tag, weigh, and measure an incredible thirty to fifty turtles that were hauled up in a single thirty-minute tow. In less than a year, more than two thousand were hauled offshore, away from the dredge's drag arm, and turned loose. Some went back to the channel and were caught again, but most others swam away (see Figure 8.3).

Figure 8.3 Loggerhead turtles on the deck of a Cape Canaveral trawler

Courtesy of Gulf Specimen Marine Laboratories, Inc.

Why the turtles aggregated in the Port Canaveral Ship Channel remained a mystery. But years later, loggerheads, ridleys, and greens were still being sucked up in the same dredge. Glen Buffkin, a Canaveral shrimper, was being paid $2,000 a night to drag away the turtles. Glen had an amazing tolerance for mobs of graduate students from the University of Georgia weighing, measuring and tagging turtles—something most shrimpers wouldn't do. Aboard the dredge, more graduate students paid for their tuition and groceries by walking around the steel grates and fishing turtle parts out of the boiling sludge. It was more a political solution than a biological one.

Paul Raymond had become a federal fish cop with a full-time, good-paying job. A slight young man, he looked more like an academic than a policeman sneaking up on commercial fishing trawlers working in closed areas at night on a Coast Guard vessel running without lights.

Coxswain Frank Makson handed us heavy bulletproof vests, and then protective exposure suits, and finally life jackets, all required for visitors. Off we waddled with the rest of the crew and Paul. He planned to interview and possibly board as many shrimp trawlers as possible and check for compliance with TED rules.

With flags flying the Coast Guard insignia and the American colors, the 45-foot vessel moved out into the Port Canaveral boat basin. We passed a dozen shrimp boats that were moored to the dock. Paul pointed to the thick orange or blue panels of heavy, large mesh webbing that looked so different from the small green webbing of the wings and tail bag. "Those are soft TEDs," he said. "The shrimpers like them, but the turtle people don't. They want them outlawed, because little dinner-plate-sized ridleys sometimes got caught in the folds of the net and drown." Several years later hard TEDs were required. They could easily be spotted from afar by their large metal hoops and excluder bars.

The shrimp boats were beautiful, graceful craft, with their nets raised and draped from the masts. But it was a bad sign for us that so many were at the dock and not out there fishing. The fleet moved with the white shrimp, following them down from the Carolinas in later summer to the northeast coast of Florida in the fall. As seawater temperatures dropped, the white shrimp ended their migrations off Central Florida. By now most had been caught or had died, and only a few boats scrapped for what was left.

We passed grain elevators, docks, parking lots, and berths for cruise ships built on the mountains of dredge spoil created forty years ago, when the port was carved out of the salt marshes and mangroves. We passed a sea of bleak metal buildings, and a hodgepodge of sailboats, tugs, and barges, and then we reached the scallop plant. Noise and stench from the rumbling machinery filled the air as a never-ending river of scallops clattered, rattled, and jiggled on the conveyor belts that carried them into the tin buildings, where clever machines steamed them and extracted their meat. The shells piled up in mounds, to be sold to developers to make driveways.

We entered the channel, cleared the jetties, went past the clanging bell buoy, and headed out into the rolling Atlantic to search for working shrimp boats. Swells crashed on the beach and broke into foam as a strong north wind came in gusts. Waves splashed over the bow as they opened the throttle.

The five Guardsmen were dressed alike, in blue uniforms with their names printed on the shirts. Paul Raymond had an NMFS insignia sewn on his jacket. All of us wore regulation orange life vests. As we stood there on deck, sweltering in our bulletproof vests and life jackets, we realized what a contrast there was between these men and commercial fishermen, with their scratched, worn-out white boots and faded clothes.

Unlike the shrimp boats, the patrol boat had no name, just a number, 41319. The crew proudly recited the vessel's specifications. It had twin 903 Cummins turbocharged diesels that could travel at 26 to 28 knots, but it had no personality; it wasn't a "her." It was a machine, designed to enforce United States laws. But it was maintained well. It had not a speck of rust because of the endless hours of chipping and painting by young men.

The crew carried a machine gun, automatic rifle, and pistols, ready to handle any situation that might arise. The words COAST GUARD were printed in big bold black letters so that there would be no confusion. As we sped south out into the rolling Atlantic, they cheerfully told us stories about recreational boaters who ran out of fuel and drifted away in the Gulf Stream, or divers who didn't set their anchors properly and came up to find that their boats had drifted off.

The seas were now running a good seven feet. The powerful engines climbed over the rolling swells and crashed through the waves, spraying water over the bow, drenching the windows. Amid exuberant shouts, Crewman Mike Ricketts boasted, "We almost rolled her over once! On a good day it's more fun than a roller coaster ride at Disney."

The patrol boat beat its way over the seas, past condominiums on an overbuilt shoreline, until the crew spotted a trawler offshore and turned toward it. Looking through binoculars, Paul told them to break it off. "That's a scalloper; looks like he's under way. I don't want to run all the way out there for nothing." Shrimp boats and scallopers look alike; they both have winches, masts, and outriggers. Only by looking at the nets they drag can you tell the difference. Scallopers and rock shrimp boats weren't required to pull TEDs, although both were known to catch turtles. At last we came upon a working shrimp trawler with a great cloud of gulls. "I recognize that boat by the black rigging," Paul said. "That's the *Captain Daddy*. Let's go talk to him."

The crew was at work on the stern, culling off; the birds were screeching and diving. The seas rose up and the seas rolled down as the white Coast Guard vessel approached, entering the cloud of seagulls. A sharp voice came over the radio: "*Captain Daddy* to the U.S. Coast Guard."

The coxswain handed the mike to Paul, who introduced himself and asked the captain what kind of nets he was pulling and whether they had TEDs. In a straightforward manner, the captain informed Paul that he was pulling Morrison TEDs, on a balloon net for white shrimp. A series of questions and answers about the catch followed. Finally Paul said, "That's David Cook; I know him. He's from Fernandina Beach. Last year I had to write him a warning because his shooter didn't open as far as it should, but he's a good guy. He uses legal gear, tries to obey the law, keeps his boat well-maintained, and catches lots of shrimp." Paul decided to move on. He spoke into the handset: "We may be back and do a boarding later, but if we don't, have a good day."

A couple of miles farther south, we came upon the *Sugar Two*, a 68-foot shrimp boat that was also dragging, and again Paul interviewed the captain on the VHF radio. He asked the captain how long he had his nets out. The skipper replied gruffly that he had just set out, and it would be a couple of hours before he hauled back. "Yeah, they'll all say that," the coxswain said grumpily. "Every one of them will say they just set out." The shrimpers knew that if the Coast Guard had to hover for any length of time, they'd get tired of bouncing around and take off.

Unless they were sinking and needed to be rescued, or a deckhand had his hand cut off by the winch, or the captain had a heart attack and had to be airlifted to a hospital, the appearance of a U.S. Coast Guard vessel was as welcome to the shrimpers as a case of smallpox. These efficient and courteous young men did their duty by the book. If they found even a minor violation, like a fire extinguisher that was out of date or a burned-out light bulb, they'd write a ticket, and the fisherman had to appear in federal court and pay a stiff fine. A serious fisheries violation meant that the Coast Guard could seize the catch and sell it at public auction to a fish house and have the check written to the U.S. Treasury. It seemed incredibly draconian. Speeding trucks on the highway got only tickets, not seizures.

The federal agencies had checked so many boats over the past year that the shrimpers had almost grown used to being boarded by men in blue with their weapons. The government had made its point. The fishermen didn't like it, but they were grudgingly accepting the fact that they had to use TEDs. As Coast Guard Cutter #41319 zoomed away, Paul said, "We try to accommodate them, and let them finish their tow, but sometimes if the weather is bad and we can't stay around, we ask them to lift their nets so we can make an inspection. We have to have probable cause. We have to catch the man towing without a TED to make a case."

Four miles farther down the coast a big paint-worn rugged shrimp boat with rusted rigging appeared. The name *Charlie G* was so faded it was hard to read. It made a striking contrast to the two other well-maintained and painted trawlers we had just passed. The *Charlie G* was under tow, and the deck was clean. A whole pack of hungry gulls were perched on the stay cables and railings, waiting for the nets to come up. Since most of the fleet had gone home for Christmas, the seabirds focused on the remaining few (see Figure 8.4).

Figure 8.4 The shrimp boat *Charlie G*

Courtesy of Gulf Specimen Marine Laboratories, Inc.

The white birds looked peaceful and settled in. Obviously, no one had come out on deck for a while to disturb them. The captain and crew seemed to be asleep in the pilothouse, because it was a long time before they noticed our presence, even though the Guardsmen hailed them on the radio three times.

With vexation rising in his voice, the coxswain gave a final "United States Coast Guard hailing the *Charlie G.*" When that failed, he blew the siren briefly. Onboard, a scruffy-looking young man opened the cabin door, looked out, saw us, and ducked back inside.

A few minutes later, a scratchy and weary voice came on the radio. "Go ahead, Coast Guard." The man sounded sleepy; obviously his deckhand had awakened him. Paul introduced himself and started the interview with the usual questions about the catch and how long he had been under tow. The captain said he had just set out. Then the fisheries officer asked him about TEDs.

"Tags?" the captain said in a puzzled tone, mishearing Paul's opening question. "Ain't got no tags." There was a pause, until the voice reflected in a low-country drawl, "Years back we caught some of them tagged shrimp in South Carolina, but I ain't seen any here." His voice sounded hoarse and tired. He was referring to a study the state had conducted that proved white shrimp migrate from north to south in the winter and back again.

"No, TEDs—turtle excluder devices. Are you pulling those?"

There was silence, and then the voice came back: "No, I'm not."

Instantly the Coast Guardsmen, who had been standing around the pilothouse looking slightly bored, perked up and crowded into the cabin, standing tensely behind us.

The shrimper's behavior was anomalous. Anything could happen. He could be a dope smuggler coming home from South America, trying to act like a shrimper. This could be one of those dangerous situations we were warned about. Aside from that business of tagged shrimp, he seemed out of touch with the real and current shrimping issues. He met all the requirements of someone whose story didn't hold together.

The Guardsmen looked at Paul with expectation. "Let's board him, tell him to haul back his nets," Paul said sharply. "The *Charlie G* is from Port Canaveral. He should know that TEDs are required here. The regs have been in effect for seventeen months. It should be common knowledge by now."

Paul turned to us. "There's no excuse," he said severely. "A local boat, and he doesn't even know what a TED is. All the other shrimpers have them." He was indignant because he had done his best to spread the word and had made every effort to help shrimpers become aware of the new law.

The coxswain pressed the handset button and barked into the radio, "Lift your nets, and stand by for boarding. Have your registration ready and the crewmen's identification. How many are onboard?"

"Three," came the sullen and defeated reply.

"And how much white shrimp?"

"Got three, maybe three and a half boxes," the unhappy voice returned.

The officer on the patrol boat was on another radio, reporting back to base so that they could run a sheet on the *Charlie G*.

"It's always the deal. Derelicts give us the most trouble and have the most violations," the coxswain muttered to us. While the Guardsmen were getting their paperwork and inspection sheets ready, the interrogation continued. The officer asked what weapons they had onboard and ordered them unloaded with the breaches open and laid on the deck. Meanwhile, the birds left the stay cables and swam all around the boat. The pelicans and seagulls didn't care about the legalities; they just wanted to be fed. They were very happy when the cabin door opened and three men shuffled out on deck and started reeling back the nets.

The crew on the *Charlie G* was a rough, unshaven group of men. Even from a distance the captain appeared grizzled and older than most men in that profession. The two crewmen were much younger, but they all looked tired or hungover. As they worked the winches, they cast worried, resigned looks at the Coast Guard vessel that hung off with its engines revving, holding its position in the rolling sea.

When the nets emerged, it was obvious that they were TEDless. Both trawls were light and popped right up; there wasn't much catch in them. The shrimper was telling the truth when he said he had just set out. But Paul said he could have been dragging a while, because in the dead of winter, when the water temperatures were nearly 50 degrees, a three-hour drag might produce only a washtub of catch.

The old vessel pitched dangerously, rocking back and forth, back and forth. The seas were running higher still. It was enough to make one seasick just looking at the men onboard clinging to the rigging for stability.

We asked if we could go aboard the shrimp boat with the Coast Guard boarding party. "Absolutely not! Boarding is a dangerous, risky time. I want you to stand back," the coxswain ordered us in no uncertain terms. "Coast Guardsmen have been shot boarding boats. The two most dangerous times are approaching the vessel and getting off. Anything can happen, and the last thing I want is you two writers in the middle of it!"

We watched from the deck as the armed Guardsmen and Paul Raymond moved to the bow. The coxswain worked the throttle, shifting into forward, and then into reverse to hold his position. Paul clutched a large briefcase full of papers, and one of the crewmen had a clipboard. As they came behind the outstretched outrigger, crewman Mike Melchoire deployed a large red crash bumper float to cushion any impact between vessels. The bow rose on a mountain of green water until they were within jumping range, but at the last second engineer Gregg Eames told them to wait while he backed away and returned at a better angle.

"Now!" he cried. The first Guardsman in line leapt, tackled the bulkhead, and scrambled over the side and onto the deck of the *Charlie G*. He helped the next man get aboard, just before the patrol boat backed away. Two men were onboard, and two to go. The shrimpers stood by on deck, watching the uniformed armed men board their boat. The crew aboard Cutter #41319 studied the shrimpers, ready to use their weapons if the need arose. But no one came out of the wheelhouse shooting. When everyone was aboard, the cutter backed away and hung off.

For 45 minutes the boarding party proceeded with the inspection, checking the catch down in the ice hold, looking for drugs, inspecting for life jackets and a fire extinguisher. The crew was held at bay on the stern and interrogated by two Guardsmen. The elderly captain led two others into the cabin to check his papers and look at the firearms. They called in the serial numbers on the guns. Almost every shrimp boat carried a rifle or shotgun to deal with net-tearing sharks.

Meanwhile, the fish and the handful of big jumbo white shrimp they had caught flopped around on the deck until they expired. All we could hear was the power plant puttering above the rumbling of *Charlie G*'s diesel. The violent sideways pitching and rolling of the shrimp boat stopped as the captain was allowed to straighten out the boat and move into a forward sea.

A radio discussion went back and forth between the Coast Guard base and the patrol boat, giving the social security numbers of the crew and registration number of the vessel to see if they were on the hot sheet. A Guardsman reported that the vessel was registered not for shrimping, but for mackerel fishing.

Other shrimpers were eavesdropping on Channel 16. "Half the boats out here say mackerel fishing on them," an angry shrimper's voice broke in. "There's nothing wrong with that, Coast Guard! Half the boats out here are registered as mackerel boats!"

Clearly this was an education process. If the NMFS wanted to make its presence known and let shrimpers know that TED regulations were being enforced, it was doing a good job. Still encased in our bulletproof vests and life jackets, we were getting hot and verging on seasickness. Finally, the welcome news came. "Ready for reboarding" crackled over the radio.

Had the blitz of pelicans and gulls that were riding the waves like ducks understood, they would have been happy as well. They were waiting impatiently for this human stuff to be over with so that the fish would get shoveled overboard. It's no easy life being a pelican. While following a shrimp boat may be easier than hunting schools of fish, they still have to wait for just the right moment and then rush in ahead of their fellows and get all the fish they can before the bycatch sinks into the depths.

With the boarding complete and the waiting over, the Coast Guard cutter eased up, trying to get their men off. It was a dangerous time. One fisheries agent smashed her knee on a boarding and had been on crutches ever since. There was a crack as the cutter's nose collided with the wood hull in spite of the pilot's best efforts to avoid it.

One man held off the shrimp boat with the big orange float. The Guardsmen helped each other as they leapt down to the deck. Then, "All men aboard," the crewman reported to the radio's command center. The patrol boat backed away, turned, and went speeding back toward the dock.

"That wasn't an easy boarding," commented Paul. "Some of those swells were running a good seven feet. I've been out there at other times when the coxswain refused to board, said it was too dangerous." As we headed back to the dock, we wondered if having writers onboard taking pictures might have had something to do with it.

Paul reported that the skipper had pleaded ignorance. The *Charlie G* had been a scalloper, and the captain had just gotten into shrimping three days before. The boat was owned by a large seafood restaurant, Captain Ed's, and had been in dry dock for two years. The fish house owner at Canaveral Seafood gave him a couple of old nets; he was trying for anything he could catch for the restaurant.

The captain knew the other boats had been using TEDs, but these old nets were the only ones the fish house had available. When Paul ordered him to stop shrimping and threatened to confiscate the catch and the vessel if they continued to operate, the gray-haired captain said he'd go in, get someone to sew in a TED, and go back out. That seemed a bit strange, since all the other shrimpers were giving it up, but shrimp were scarce, he needed the money. Christmas was coming, and the restaurant needed seafood badly.

Paul decided not to confiscate and sell the catch, although he had every legal right to do so under the Magnuson Fisheries Act. "There will be a fine, but the amount will be up to the judge. When I said it could be a $5,000 fine, he just shrugged. That would have waked me up, but he didn't seem to care."

When the *Charlie G* came back to the dock, the captain blew town and was never seen again. Captain Ed's Seafood Restaurant threw itself on the mercy of the court. Mercy was in short supply, and the owner of the boat and the fish house were fined $8,000 apiece. It was the first case ever made on TEDs. One small skirmish in the long battle between shrimpers and conservationists, it was a hollow victory that pitted conservationists trying to preserve the planet's biodiversity against commercial fishermen trying to preserve their tough, traditional way of life.

Eventually, shrimpers, who are always tinkering with their nets, perfected the TEDs and reduced shrimp loss to 10 percent. In recent years Kemp's ridleys have made a dramatic comeback, and greens, loggerheads, and hawksbills have become more common in U.S. waters.

During hearings on the reauthorization of the Endangered Species Act, Congressman James Traficant of Ohio complained, "We have spent more time on turtles than we did on Star Wars, we have spent more time on turtles than we did on jobs, and we have spent more time on turtles than we did on the American farmers and coal miners." He was not amused. "Let's get on with it."

The fishing industry's appeal in federal court to block the use of TEDs was eventually lost, as were similar suits against TEDs in South Carolina. Finally, in 1990, American shrimpers were required to install the panels that would shoot turtles out of their nets. The turtle populations began the slow road to recovery, and the shrimpers continued on their downward spiral to near economic extinction.

While American shrimpers had to use TEDs, foreign fleets had no such restrictions. Congress passed legislation requiring imported shrimp to be caught with TEDs. Even though leatherbacks were going extinct locally, and olive ridleys were rapidly declining from overharvest of eggs, Malaysia, Thailand, and India sued in world court to prevent the U.S. government from requiring them to use TEDs if they wanted to sell shrimp in the U.S. Exporters sued successfully in World Court, and again American shrimpers lost.

Even where governments agreed to require TEDs, enforcing the regulations worldwide was a difficult matter for NOAA and the State Department. Honest impromptu inspections or surprise visits were not allowed. Notice was always required, and U.S. Fisheries officers had to be accompanied by local officials when they came down to the docks. With several weeks' notice, shrimpers had plenty of time to add a TED to the net for the occasion and pretend to be in compliance.

In 1991, when I (Jack) went to Malaysia to learn how to process jellyfish for food in an effort to help beleaguered American shrimpers, I visited a busy modern shrimp-processing plant in Thailand. A huge fleet of brightly colored trawlers were docked along the Saluween River, which separated Thailand from Burma. Rivers of shrimp were carried into the factory on clanking conveyor belts, where women and children headed, cooked, peeled, and deveined them. They were breaded, packaged, and frozen.

None of the boats had TEDs in their nets, but many had machine guns mounted on their bows, with swarthy, rough, and dangerous-looking men with machetes and sidearms glowering suspiciously at me. The owner graciously invited me and the Chinese jellyfish merchants who were hosting me to a tea ceremony. While we were talking about conservation issues, a captain came and turned in a machine gun and a belt full of bullets. TEDs, said the plant owner, were the least of his problems. Having their catches taken at gunpoint, having his crews murdered, and having his boats stolen by pirates were far greater concerns.

Even though the U.S. government held workshops and training in Costa Rica, shrimpers still spurn the use of TEDs. In the past four years, the National Coast Guard caught twenty-nine boats cheating on TED regulations, yet not a single sanction was imposed. The U.S. State Department levied temporary trade embargoes to force compliance, but violations continue. On the other hand, at least some shrimpers in Madagascar voluntarily embraced TEDs, saying they cut down on bycatch and work.

Today, even with 98 percent compliance by the American shrimp industry, sea turtles continue to wash up on the beaches from time to time. Hence, the environmentalists have pressed for more efficient TEDs. When a number of leatherbacks were stranded, the government forced the shrimpers to expand the openings to 71 inches so that the rare behemoths would be shot out, along with more shrimp. At the time, there were a large number of these smooth-skinned black-and-white monsters that weighed up to a thousand pounds, passing through the East Coast waters eating cannonball jellyfish, but the new size holes became mandated Gulf- and Atlantic-wide.

Steve Kirchner, a spokesman for the shrimping industry in South Carolina, expressed the shrimpers' frustration: "No matter what happens, they blame the shrimpers. You talk to officials and see the glaze come over their eyes. You have to get through by saying something different than 'You're destroying our livelihood with your regulation.' They don't listen to that. They hear it all the time, and they don't care."

Sea turtles were only the beginning of the shrimp industry's environmental problems. Environmental groups publicized the fact that for every pound of shrimp that reaches the table, nine to fifteen pounds of other dead and dying fish and invertebrates were shoveled over the side. No longer did the public view the romantic shrimp boats putting out to sea as a wonderful way of life. Suddenly they were seen as death mills that churned up and destroyed the bottom and left a wake of dead fish and sea turtles.

Life expires quickly on the deck of a shrimp boat when it's avalanched and crushed in the alien world of air. Seeing the eyes glaze over and life slip away is not easy. Squid and anchovies die quickly, while others, like stingrays, linger. In the wintertime, many of the fish make it back alive, but in the broiling heat of the summer, probably very few survive.

Shrimpers say they're feeding the ocean; environmentalists counter they're disrupting the normal behavior of marine life and destroying the world. The National Marine Fisheries Service created a furor when it declared that red snapper was overfished in the western Gulf of Mexico and imposed quotas on both commercial and recreational fishermen in 1997. Recreational fishermen were infuriated when catch limits in federal waters were dropped again to two fish per boat in 2007. A barrage of court challenges from recreational and commercial red-snapper fishermen followed, claiming that the source of the problem was shrimpers who killed juvenile fish in their bycatch.

In the 1990s, a growing environmental awareness developed. As nature films and programs became part of regular television programming, people watched beautiful reef fish swimming across their television screens like aquariums. Reef aquariums became commonplace in homes. Reef-tank technology improved, pet shops proliferated, and aquarium enthusiasts could watch dazzling reef fishes and colorful invertebrates in their homes. International ecotourism increased, the diving industry mushroomed, and people swam, snorkeled, and dove on the magnificent undersea coral gardens in the Bahamas, Belize, and far-away places like the Philippines, the Seychelles, and the Red Sea. Conservation groups proliferated; gill nets were banned in Florida, and shrimp nets were limited in size in inshore waters. Across the globe, and particularly in the United States, awareness spread that juveniles of valuable fish species were being killed in the process, including trout, croaker, drum, sharks, and snapper.

In the central and western Gulf of Mexico, the valuable red-snapper fishery had declined catastrophically, and shrimpers killing juvenile fish were the probable cause. Off Texas and Louisiana, in tow after tow, hundreds of the little orange fish that could have grown up and provided marvelous meals to people were shoveled over the side dead along with the rest of the "trash." Commercial hook-and-liners joined the recreational fishing organizations in demanding that the government put an end to the overcapitalized and oversized shrimping industry and do it quickly. The men who dragged the sea became scapegoats for all the factors affecting the loss of productivity in the sea.

Well-financed sports fishermen began a smear campaign, labeling shrimpers "bikers on boats" and "resource pirates" who raped the seas. TEDs were soon followed by bycatch reduction devices (BRDs). Regulations, closures, and restrictions rained down upon the shrimpers' heads.

The new political reality was expressed by the website www.Shrimpsuck.org: "Shrimp is the number-one seafood in the U.S. Eating shrimp can be among the worst things you can do to the ocean. Bottom trawling and farming for shrimp can destroy the ocean. You don't want to be a part of that!"

Sports fishermen, who were trying to put an end to shrimping, labeled bycatch waste, waste, waste! Yet even with a hundred boats working the area, the billions of fish and other creatures that perished on the decks were instantly assimilated by the sea. Most of the time, no fresh dead fish came up in the trawls. It didn't take a great stretch of the imagination to guess what happened to the bycatch. There are winners and losers at all levels of shrimping. The flocks of seagulls, jack crevalles, porpoises, and sharks gulped down the dead and dying as fast as the leftovers were raked overboard. And down below, crabs and even loggerheads and Kemp's ridley sea turtles feasted on the sudden flood of protein discharged into the system. However, on rare occasions, when the tide is slack and hundreds of boats are working a small area in the summer heat, the discards of a hundred trawlers or more working an area pile up on the bottom and rot. The bottom sours, the shrimp vanish, and the shrimpers move on.

A necropsy performed on a sea turtle that washed up on a Cumberland Island, Georgia beach had jumbo shrimp, squid, and fish in its stomach—all fast swimmers that a turtle could never normally catch. This raised the possibility that some loggerheads were swimming into the net to partake of the catch and then turning around and escaping. Meanwhile, another feeding frenzy was taking place on the bottom: blue crabs, catfish, trout, and the shrimp themselves were having a feast.

Even though mackerel, flounder, and whiting were also in decline, Congress didn't want to hurt fisheries' biggest cash cow: shrimp. The annual crop of 350 million pounds dragged up from the waters of the Gulf of Mexico and Southeast brought $450 to $500 million into the economy. Whether the conservationists liked it or not, whole towns and economies depended on shrimp. If the industry folded, in addition to the fishermen, a giant cadre of harvesters, processors, wholesale and retail distributors, and truckers would be put out of work. Behind them would come a secondary economic collapse as boatyards, maintenance shops, gear, fuel, and ice docks closed their doors.

After mandating the use of TEDs, Congress ordered the National Marine Fisheries Service to do something about bycatch for all fisheries that entangled, hooked, or killed seabirds, sea turtles, seals, dolphins, and even whales in their fishing efforts. The NMFS's fisheries engineers came up with BRDs, which were panels or devices that were inserted into the trawls to let the fish out.

Seeing that legislation would pass, the shrimpers again allowed observers to ride their boats and measure the catches and assisted them in developing BRDs, hoping they wouldn't lose too many shrimp. And if they did succeed, it would mean less bycatch and less work culling it off. From 1993 to 1996, NMFS observers logged 5,700 hours using nets with and without twelve different types of BRDs. They weighed and measured the catches from all over the Gulf of Mexico, the

Tortugas Fishing Grounds, and the South Atlantic. They tested them on white shrimp, rock shrimp, pink shrimp, and browns, comparing the catches of nets with and without BRDs. Never in the history of fishing were catches given such scrutiny. Of the 60-pound average catch per hour, the observers found that only 16 percent of the total catch by weight was shrimp. The other 84 percent included more than 450 species in the Gulf and 150 in the South Atlantic.

Eventually the NMFS came up with several designs that reduced the bycatch by 30 percent, but this fell short of the mark of achieving a 50 percent reduction in the juvenile red snapper catch. Croakers, spots, mackerel, spotted sea trout, and other finfish swam out of the openings, but not nearly as much as the conservation community wanted. Studies showed that 37 species of crustacea, 166 fish, and 29 other invertebrate groups were not excluded from nets. Nevertheless, the shrimpers were required to use both TEDs and BRDs, and they got stiff fines if they didn't.

When the red snapper population failed to recover, the Gulf Coast Conservation Association, a powerful group of recreational fishermen, smelled blood. They filed suit in federal court, demanding an end to trawl fishing. They were joined by some prominent and well-publicized marine biologists who did the lecture circuit and appeared on public radio and television, predicting that because of the worldwide trawl fishery and its horrendous impact on marine life, seafood would become a thing of the past in the next fifty years.

Over the years, shrimpers have altered the seafloor and damaged fragile biological communities with their nets' chains, foot ropes, and heavy iron-clad doors dragging up all that sponge and life. They have drawn the ire of the environmental community and conservationists. However, not all trawling is destructive, especially when nets are dragged over sand or mud bottoms. Dr. Joseph Donohue, a coastal geologist working in the Florida Panhandle, told us that during a strong northwestern blow, up to three feet of sediment from the Gulf seafloor was churned up by wave action, which was far more than any shrimp net did. Furthermore, during tropical storms, six feet of sand and mud could be stripped away. But that nuance was lost on the general public.

As the fleet overexpanded in the 1980s and 1990s, a few fishermen desperate to make payments on their newly financed boats moved into new areas that had been considered too rocky to drag nets on, and they severely damaged the bottom. Some captains took their nets off the doors, added heavy chain to the doors, and dragged the shallow bottoms, deliberately snapping off coral heads. When the bottom was flattened, and the coral heads and sponges were broken from their holdfasts, the nets were put back on, and they caught tons of pink shrimp that were hiding in the structure. Shrimpers often talked about "breaking in new bottom" and likened it to farmers plowing a field so that they could plant crops.

But some stony corals take decades or even centuries to grow. In the 1970s, we were contractors working on a fishery management plan for coral. To fulfill bureaucratic requirements, we were hired to describe the "coral fishery" even though there wasn't one. Harvesting coral in Florida waters had been illegal for years, but the government wanted to show it had covered all the bases. We could have said "no coral fishery; that's that; let's go home." But when we used the opportunity to blow the whistle on the trawl damage to corals that was occurring in the Apalachee Bay off St. Marks, Florida, we were summarily fired. The highly political fisheries councils were dominated by fisheries lobbyists who didn't want to hear it. The damage couldn't be covered up for long, however. Trawling with door nets was eventually prohibited in that area. The coral heads and gorgonians are now thicker than ever.

The conflict over trawls is nothing new. In 14th century England, during the reign of Edward III, a petition was presented to Parliament in 1376 calling for the prohibition of a "subtly contrived instrument called the wondyrchoum." This was an early trawl with a wooden beam and a net 6 meters (18 feet) long and 3 meters (10 feet) wide.

The petition read:

> "*...in several places within your said realm, in creeks and havens of the sea, where was accustomed before these times to be a good and plenteous fishery, to the great profit of the realm, which is in part destroyed and rendered useless for a long time to come by some fishermen who have...contrived a new instrument which is amongst themselves called a wondyrchoum...to which instrument is attached a net of so small a mesh, no manner of fish, however small, entering within it, can pass out and is compelled to remain therein and be taken; and besides this...destroys the spawn and brood of fish beneath the said waters and also destroys the spat of oysters and mussels and other fish, by which large fish are accustomed to live and be supported...the fishermen...take so great abundance of fish...that they*

know not what to do with them but feed and fatten their pigs...to the great damage of the whole commons of the Kingdom and the destruction of the fisheries in like places, for which they pray remedy."

The arguments remain the same then and now. Governments have struggled with protecting resources for centuries, closing bays and sounds periodically to allow shrimp and other marine life to grow larger. As a member of the Florida Marine Fisheries Council said, "We can't manage fish; we can only manage fishermen." In 1981 the Cooperative Texas Closure stopped shrimping for two months out of the year for 200 miles out to give brown shrimp, the most valuable fishery in the nation, a rest. That same year, nearly 5,000 acres was permanently set aside to create the Tortugas Pink Shrimp Sanctuary.

The Gulf of Mexico and South Atlantic Federal Shrimp Fisheries Councils, established under the Magnuson-Stevens Fisheries Act, created Exclusive Economic Zones and mandated U.S. Coast Guard enforcement of their regulations. The Councils then set limits and quotas, regulated gear and mesh size, and implemented further TED and bycatch regulations. By 2002 any boat shrimping in federal waters was required to have a permit, and shrimping became a limited-entry fishery when the government placed a moratorium on new fishing permits. Because of these actions, there are more shrimp than ever. Over the past thirty years the collective landings for pink, brown, and white shrimp have remained the same—around 150 million pounds a year. But since 2002, possibly due to the expanding dead zone, collective landings have dropped to 92 million pounds as the fleet has downsized.

9

Wetlands and Real Estate

The biggest problem with managing the marsh and mangrove wetlands that are the nursery for juvenile shrimp is that managers cannot always protect wetlands from pollution or politically well-connected real estate developers. It's easy to beat up on the small commercial fishermen, but land use is controlled by wealthy, powerful politicians and real estate developers, some of whom are political appointees sitting on the fisheries councils. Many freshwater wetlands, in particular, have been stripped of much of their previous legal protection by recent Supreme Court decisions, contributing to degraded water quality in coastal wetlands.

It's been shown again and again that when land speculators move in, wetlands and shrimp diminish, and fisheries collapse. As a study in Texas showed, when developers seal off the land from the sea with seawalls, they rob the shrimp and other marine life of organic materials and diatoms that occur in the marsh-mangrove interface. Then shrimp landings drop catastrophically. Two and a half times as many brown shrimp and fourteen times more white shrimp were found along a natural shoreline than along a bulkheaded shoreline in one study. Another scientist at the U.S. Fish and Wildlife Service stated that the loss of organic material along bulkheads in Florida, Louisiana, and Texas dropped shrimp to one-eighth of their normal population. Dams, causeways, impoundments, and storm-water drains all cut down on shrimp population, according to USFWS studies. But all the scientists did was study it. The prevailing trend has been that politics and money trump science. The unrelenting beat of "dredge, drain, and reclaim" still goes on to create the real estate dream, a bulkheaded seawall and a manicured grass lawn. People want to "improve" their property and make it look clean and neat, not wild, so that they can boost their real estate values.

Nearly half the wetlands in the United States have been destroyed by dredging and filling over the past century. Jeret Madei, an artist, environmental activist, and old friend, grew up in Fort Walton Beach, Florida in the 1950s and watched the Intercoastal Waterway being built. Fort Walton Beach was hardly a town. It was mostly an Air Force base built on a little sandy peninsula, surrounded by salt marsh.

"Between 1953 and 1957 you could look out over Choctawhatchee Bay, and everywhere you could see huge areas where the white shrimp were boiling at the surface," Jeret recalled over a shrimp dinner one night at our house. "There were birds diving on the shrimp, fish striking, and dolphins everywhere. All Fort Walton was a strip of barrier beach, with giant marshlands and swamp behind it, and the swamps went way inland, all the way up the rivers, and made a giant nursery ground for the shrimp.

"We'd go to the Brooks Bridge Docks and wait for the fishing boats to come in. The boats had big scales hanging on the deck, and they'd weigh out the shrimp, but we had to bring our own ice. We could buy the shrimp for five cents a pound, raw with their heads on, or pay more for cooked. Some had their cookers onboard and would sell boiled shrimp for ten cents a pound. The mosquitoes were fierce. So many were swarming, you breathed them in; the night air was thick with them. They'd plaster the windshield when you'd drive, and when we went across the bridge, we had to keep the windows rolled up even though it was hot. They'd lay eggs in the mud, and the water was filled with larvae. Maybe the shrimp ate them."

My friend was right about shrimp eating mosquitoes. We fed wigglers to post-larval white shrimp and watched the half-inch-long crustaceans go wild eating them. Mullet and other fish are also known to eat mosquito larvae, but not enough to provide enough mosquito control for human comfort.

Mosquitoes have shaped Florida's history. Early explorers described burying themselves under the sand to keep from being eaten alive. Mosquitoes killed hundreds of people from Pensacola to Fernandina Beach and down to the Florida Keys by spreading malaria and yellow fever. Along the Carolina and Georgia Coasts, pre-Civil War plantation owners fled to the outer Sea Islands to escape them, leaving their slaves to fend for themselves. If Florida and the rest of the coastal south were to grow, those torturous mosquitoes, yellow flies, and sand gnats had to go. And go they did, with intensive ditching, draining, and land clearing. To increase development, improve shipping, and bolster the economy, the U.S. Army Corps of Engineers built the Intercoastal Waterway. They dredged up the marshes, cut off the drainage, and filled in the swamps until people could walk around in shirtsleeves and not be eaten alive.

"It was amazing how fast they did it," recalled Jeret. "In just three or four years, they filled in all that land where the shrimp docks used to be, built strip malls, condos, and restaurants that now serve farmed shrimp from China."

Long before development spread to the remote Florida panhandle, land speculators were at work chopping up South Florida's wetlands. Henry Flagler had drained thousands of acres of swamps building his east coast railroad in the early 20th century. The land boom started in the '20s, and flocks of real estate speculators wearing pinstriped suits sold swamps to suckers. As the big hotels began to rise on the coast of Florida and parts of the Gulf Coast, mosquitoes feasted on the flesh of Yankees as they got off the train. Mosquito-control programs were established and continued the war on bugs.

During the depression, the Works Project Administration dynamited and dug fifteen hundred miles of ditches through Florida's salt marshes. After World War II, spray trucks roared up and down the dikes and coastal highways, and crop dusters flew overhead, dousing the marshes with DDT and killing the bugs. But they also killed juvenile white and brown shrimp, blue crab, mullet, croaker, flounder, and other seafood, as well as hordes of fiddler crabs and periwinkle snails, which few people cared about.

According to the U.S. Fish and Wildlife Service, the lower 48 states had 215 million acres of wetlands when European colonists first came to America. The cities of Washington, D.C., Baltimore, New York, Philadelphia, Charleston, and New Orleans were built in the swamps and floodplains, as were innumerable smaller towns, airports, and harbor facilities. Land was farmed and channels were dredged until now only 105.5 million acres remain, an area about the size of California. The years from the mid-1950s to the mid-1970s were a time of major wetland loss. The wetlands that produced both white and brown shrimp were filled in, seawalls were put down, and more than half of the habitats were diminished. In Florida, sea-grass beds disappeared from sprawling coastal development. Polluted storm-water runoff and channel dredging turned underwater meadows into unproductive muddy bottoms. Although there has been much rhetoric about preserving wetlands in the last decade, the U.S. Fish and Wildlife Service reported in 2009 that wetlands continue to decline. The Gulf of Mexico lost nearly 371,000 acres, while the marshes and swamps of the Atlantic Coast went down by nearly 15,000 acres, all from human population increases and development.

As development increased, and tens of thousands of recreational vessels crammed into the bays, sea grasses were torn up by inexperienced weekend recreational fishermen. They ran their overpowered boats aground in the shallow grass beds and dredged their way back to deeper water, churning up the bottom and creating long linear swaths of bare sand called prop scars as they went. Once the rhizomes of turtle grass are exposed to oxygen, the roots die off, creating prop scars that take decades to grow back.

Florida led the nation in obliterating its wetlands. One study by the state, using global satellite positioning, showed that 86 percent of the mangroves from the Indian River to St. Lucie Inlet are gone. Developers in Tampa Bay on the Gulf Coast gobbled up 81 percent of the marshes and mangroves. Jacksonville chewed up over 12 miles of riverside marshes of the state's largest estuary to make way for shipping ports, housing, and commerce. It's estimated that 78 percent of Florida's residents now live in coastal areas and former wetlands, and fish are confined to the undeveloped remnants. Seawalls and buildings have replaced mangrove swamps, depriving shrimp and other marine life-forms of food and shelter, giving them no place to grow.

Even the people who make their living from the sea often don't get it. Pompano fishermen on the Florida east coast took jobs destroying the marshes of Lake Worth, working for the developers when fishing was poor. After the marshes and mangroves literally went down the tubes (dredge pipelines), the flounder, pompano, mackerel, whiting, crabs, and shrimp vanished, and the fishing industry collapsed.

A few commercial fishermen still struggle, driving their boats up and down the highways and launching at public boat ramps, trying to make a living from the sea. They're reminiscent of a flock of egrets we once saw in Chiapas, on the Pacific Coast of Mexico. Behind a chain-link fence, in an industrial port, the harbor was being expanded. Behind the newly built seawall were a couple of acres of stagnant marsh that had been cut off from the sea. The birds were fighting over the last fish in the last mud puddle that was being filled in. A few hundred feet offshore the giant dredge was at work, deepening the channel, and pumping the spoil into that last remnant of marsh.

While people argued about the impacts of dredging in Florida, it wasn't even a consideration in Louisiana. The U.S. Army Corps of Engineers has been building levees and trying to straighten out the unruly Mississippi River for decades. In 1927 the river flooded 25,000 square miles, and 637,000 people were forced to flee their homes. Perish the thought of not building in the floodplain. The Corps stepped up to the plate and built higher levees with concrete retaining walls and spillways to divert the river and protect the deepwater port of New Orleans. It's possible to drive along the highway parallel to levees near Morgan City and see tugboats pushing barges above your head. To aid the exploration for oil and natural gas, dredges and draglines have sliced and diced the Louisiana swamps for the past hundred years, sometimes removing a million cubic feet of sediment at a time. As offshore oil fields were developed in the 20th century, the drilling barges needed canals to float. Starting in 1937, crawfish and catfish fishermen had to get out of the way as the cypress swamps and marshes were chopped into a grid of canals.

The only problem is that channelization has robbed the delta of land-building silt, carrying it offshore and dumping it off the continental shelf. Without the river replenishing it, the land is rapidly eroding and washing away. The recent increase in frequency and intensity of hurricanes has only added to the problem.

In recent years the rapid demise of Louisiana's wetlands has caught the public's attention. With oil pipelines cutting through the marsh and development sucking out the ground water, salt water is intruding as freshwater wetlands subside. Trees and marsh grass that once provided leaves and nutrients to feed the estuaries are falling to the sea as barrier islands shrink. With them go the commercial and recreational fishing industries, duck hunting, and tourism that bring billions of dollars into Louisiana's economy.

The mighty Mississippi River that drains the entire eastern seaboard is responsible for making the Gulf of Mexico one of the richest fishery grounds in the world. Over the centuries, it has repeatedly changed course, cutting new channels and creating oxbows. It has left the old river channels to turn into backwater swamps and has built wetlands in the delta with its sediments. About a hundred and fifty years ago, Midwestern farmers began clearing the landscape and eroding the soil. The soil washed into the river as silt, so much so that it's been estimated that for every bushel of corn that's produced, two bushes of topsoil are washed away. As farmers tilled away, the downstream waters became muddier and muddier, carrying an average of 436,000 tons of sediment a day. As engineers tamed the unruly river with over a hundred dams, levees, and erosion-control structures that kept the river confined to its channel, the delta marshes were starved of sediment, causing them to erode into the sea. No longer free to alter its course, and with no way to inundate and flood and renourish the vast freshwater wetlands, the Mississippi simply shot its sediment loads into the deep waters of the Gulf instead of building up the delta.

Both brown and white shrimp share the same shallow muddy bay habitats in their post-larval stages. The Louisiana delta may be washing away and hurting the inshore white shrimp populations, but the mud pouring out into the deeper waters of the Gulf may be benefiting the offshore adult brown shrimp populations at the same time. Because shrimp love mud, it is possible that human environmental alterations on the northwest coast of Madagascar that caused massive erosion may have increased the shrimp population. For the past hundred years the mountains of the big island off Africa have been stripped of their forests and grazed to such an extent that astronauts looking down from orbit have declared that Madagascar was "bleeding" into the sea. Years of cyclones and torrential rains have washed huge plumes of red mud into the sea and smothered the bottoms. Yet, after the French research vessel *Ambariaka* discovered shrimp near Nosy Be in 1964, shrimp landings have continued to increase.

In America, man has changed the water regimes with his progress. With his ditching, channelizing, and draining the creeks for mosquito control and real estate development, freshwater no longer trickles through swamps and vegetation in many places. When it rains, the water-management agencies in South Florida open the floodgates, and water surges through the ditches and blasts out into the estuaries. Too much water delivered at the wrong time shocks the delicate physiology of many forms of marine life and over time causes it to diminish.

Hurricanes are also major agents of change for both shrimp and shrimpers. Even though it looks big from the deck of a trawler, the Gulf of Mexico is a small, nearly landlocked ocean surrounded by the United States, Mexico, and Cuba. It has just a narrow opening into the Atlantic through the Straits of Yucatan and Straits of Florida. Only during its frequent hurricanes and storms does one see giant rolling waves crashing onto shore. A good shrimper isn't afraid of bad weather, but he respects it. Boats sink and men are lost at sea when a big one hits.

In 1969 Hurricane Camille, a small tight hurricane with 200-mile-an-hour winds, slammed into the Mississippi Gulf Coast, killing 255 people and demolishing whole fleets of shrimp boats. Shrimpers who survived it were catching pipes, toilets, lamps, and furniture in their nets for months afterward. David Crum, a friend of ours, dragged up a red pony, and there were stories of human bodies being dumped out on deck. One captain rode his 65-foot double rig, the *Sandy O*, up the Pascagoula River, trying to outrun the storm surge. He and his crew would have been killed, but at the last minute, they leaped from the cabin roof onto a highway bridge before their boat was smashed to pieces. Other shrimpers drowned, and some of their bodies were never recovered.

Hurricane Camille was just a dress rehearsal for the much larger Hurricanes Katrina and Rita in 2005, when flooding and winds killed seventeen hundred people and almost destroyed the fishing industry. The shrimp industry in Louisiana, Mississippi, and Alabama suffered a nearly $2 billion loss that year. As many as five thousand commercial fishing boats were destroyed. Most of them had no insurance because it cost too much to start with, and, after Katrina, premiums jumped another 50 percent. Flooding destroyed fishing infrastructure, from docks to canneries and processing equipment, putting people out of jobs. The water became foul and polluted from human and industrial waste. Even a year later, the area still looked as if it had been carpet-bombed. If it weren't for the efforts of a few big processors that kept the industry afloat, shrimping there would have been destroyed.

One night a few weeks after a hurricane had passed, I (Jack) narrowly missed getting killed by a falling tree, twenty miles offshore. The nets came up from eighty feet of water with a full-sized pine tree wrapped up in the tickler chain. It was stuck in the webbing, and there was no way to cut it loose, so the deckhand winched the net and the attached tree to the top of the mast. For a moment it stood there, a tree on a shrimp boat, needles, branches, and all, a strange sight to behold. I didn't have long to contemplate it before gravity took over. Before anyone could cry "Timber!" it came crashing down on deck, scraping me with its branches. But the crew also shook big pink shrimp out of the reptilian-like scales of waterlogged palmetto bushes that were also hanging in the nets. The shrimp loved the vegetation because it provided both structure and food.

Bumper shrimping years often come on the heels of a hurricane. The same raging storm that sinks some trawlers and washes up others on the beach also churns up nutrients in coastal wetlands. This provides plenty of food that produces an explosive new crop of shrimp the following year. Hurricanes giveth, and hurricanes taketh away. Nutrients, trees, and vegetation are pushed into the estuaries, and shrimp eat them. Over the years this phenomenon has been reported along the coast from North Carolina to Texas. In the movie *Forrest Gump*, the entire Alabama fleet is washed up on the hill, and when Forrest goes back out and hauls in his nets, he's buried in shrimp. The story isn't all fiction; this effect is now called the "Bubba Gump Phenomenon." Some of the best shrimping comes after a big storm that shakes things up and provides brand-new structure. Waves and currents put shrimp and fish on the move. It happened with Hurricane Floyd, which blasted North Carolina in 1999 with 130-mile-an-hour winds and torrential rains. As with the storms that ravaged the Gulf Coast, people were trapped on their roofs, and lives were lost as roads were washed out that might have been used to escape. Sewage-laden floodwaters rose and flowed into agricultural and livestock farms, drowning millions of chickens and hundreds of thousands of pigs, leaving behind wreckage, filth, and destruction.

State officials predicted that, given the amount of sediment and sewage flowing through the waterways toward the coast, massive fish kills would occur in Pamlico and Albemarle Sounds, and the shrimp crops would be wiped out. Instead, North Carolina had its best shrimp season ever. Likewise, when Hurricanes Fran and Bonnie pummeled the Atlantic Coast, shrimp and crab landings increased. Even fighting the garbage and debris, the shrimpers who survived Katrina and Rita caught so many shrimp they couldn't transport them all. But few processing houses were left to unload the catches. It was almost impossible to buy fuel and ice, which thinned out the fleet. Shrimp had to be transported by truck to the few surviving fish houses. On the other hand, if the storm hits during a run, it can churn up a productive bottom, sweep the shrimp out to sea, and ruin a good season. Such is the life of a shrimper.

While hurricanes can be good for wild shrimp populations, moving them from one region and depositing them in another, they can be devastating to shrimp farms, and sometimes to the populations of wild shrimp that live near them. Not only do floodwaters overtop and breach the farm's dikes, setting the crop free to wander, the storms can also spread disease. In Honduras, when Hurricane Mitch struck with devastating fury, floodwaters laced with antibiotics, pesticides, and herbicides from shrimp farms receded into the estuaries and stressed the wild shrimp populations. The shrimp that survived the sediments and decomposing organic material had compromised immune systems. Shortly afterward the deadly white spot virus spread from Ecuador's shrimp farms to wild shrimp at sea and was met with little resistance. The already-reduced shrimp population was all but wiped out. It has taken nearly a decade for the wild shrimp population to rebound. As sea level is rising and water temperatures increase, hurricanes are growing in size and ferocity. In recent years hurricanes have become so numerous along the Gulf Coast and South Atlantic that it's hard to remember their names.

Hurricanes, overfishing, pollution, and loss of wetlands did not kill the shrimping industry. Neither did TEDs and BRDs, as the shrimpers had predicted they would. But rising fuel prices and declining shrimp prices caused by imported farmed shrimp flooding U.S. markets nearly did close it down.

Starting in the 1980s, millions of pounds arrived in freezer ships, and refrigerated warehouses were bursting at the seams. When shrimpers unloaded their boats at the fish house, they found themselves in a price time warp, getting the same price they got in 1964! For a while you could buy shrimp at Costco for two bucks a pound, and Wal-Mart ran specials on ten-pound bags for even less. Even jumbo shrimp, which once fetched nineteen dollars a pound, were on sale for six. By the early 21st century 88 percent of all shrimp sold in the United States came from foreign countries.

Meanwhile, the costs of fuel and ice skyrocketed. Owners could no longer afford to maintain their vessels and keep them in good repair. Previously well-maintained vessels began to run down. Captains, crews, and seafood workers were forced to take other jobs, hoping for better times. Many put their boats up for sale, but there were no buyers. Some hauled drugs from Colombia and went to jail. Others just dried up and blew away. By the turn of the 21st century, the fleet had shrunk to half its original size, and most of the processing plants had closed. Today even Forrest Gump would think twice before buying a shrimp boat.

Steve Kirchner, a fifty-year-old shrimper and spokesman for the shrimp industry in St. Helena, South Carolina, invited us to come down to his boat to discuss the problems that shrimpers are facing. On a cold spring morning in 2007, we drove to a remote landing, where a salt marsh house sat on a tidal creek.

Although condos and developments are proliferating elsewhere, deep in South Carolina's "pluff mud" coast, a few rickety old docks and shacks still perch on the banks of green winding creeks. Although it's uncertain where the name came from, every year people in the Low Country hold a festival to celebrate the smelly muck, wearing T-shirts emblazoned with "Pluff Mud keeps the Condos away." Combined with gnats, yellow flies, water moccasins, and alligators, pluff mud does just that.

Alligators coated with black sticky mud slide off the banks and splash into the creeks. Great blue herons and egrets, wood storks and ibises stalk the flocculent slopes of the creeks, looking for shrimp, trying not to sink in, and keeping a wary eye out for gators. Miles from the sea, deep in the labyrinth of salty green marshes, white shrimp can be so thick that they jump and skip over the water when disturbed. One shrimp touches another, and the agitation spreads until they explode like popcorn, land on the muddy banks, and thrash furiously to get back into the water before the avian cloud of seagulls swoops down on them.

We drove slowly down remote back roads trying to follow Steve's directions until we arrived at the old fish house that was out of business or closed down for the season. The tables were empty, the freezer door was open, and the concrete floor was cracked and filled with leaves that had blown in. We made our way carefully out on the rickety old dock, stepping over missing planks and walking along catwalks that spanned missing sections, battered from the last hurricane.

Having battled TEDs, BRDs, and shrimp farming for decades, Steve lectured in a folksy but articulate way to civic groups, legislators, politicians, and whoever would listen to him. His worn pants, flannel shirt, white boots, and calloused hands clearly marked him as a fisherman as he began to tell us the story he had obviously told many times before.

"We can't afford to maintain our boats anymore," Steve said morosely. "This is just a remnant of what used to be here. This steel hull is going to Florida, where it will be used as a tug for oil barges. The rigging will be taken off. Other boats haven't left the dock in two years. We were getting by on retail shrimp for a while, but now they can't compete with dollar-a-pound wholesale prices. At the height of the energy crisis, shrimpers were being forced to pay six and seven dollars a gallon for diesel fuel. Nine out of ten ended up tied up to the dock, unable to make a trip or make payments. The banks took their boats and are sending them to Alabama, where they're being stripped down and turned into longline boats, or they ship them to the Bahamas or Bermuda for freighters."

Many a trawler has settled on the muddy bottom and lies there overgrown with barnacles, providing structure and food for the little crustaceans it once sought. A grouper looks out from the door of the wheelhouse, and a moray eel hides under the rusted winch as a shark swims overhead. Other fishermen hang on, trying to sell their shrimp in little retail markets, often resentful of those who farm the sea, knowing they are losing the economic war and their way of life as the price of shrimp gets cheaper and cheaper.

Finding a reasonably priced place to tie up a shrimp boat has become a serious problem. It's rare that a marina will let shrimpers dock there. No one wants the rough-looking, rowdy, and often intoxicated crews and often smelly trawlers moored next to fancy yachts and pleasure craft. Charleston, South Carolina used to have shrimp boats everywhere along the docks, but now endless sailboats fill giant marinas. No shrimp boats can be seen, except in backwaters. In Hilton Head, South Carolina, some of them are a tourist attraction. Even finding an industrial dock to tie up to is getting harder. There was a great uproar in Galveston, Texas when the Port Authority expanded the port and pushed out the shrimpers. They dredged out the harbor and rebuilt huge industrial docks to accommodate more grain elevators and deep-water ships and seagoing barges. After a long squabble, with historians and townspeople protesting, the shrimpers were forced to move to another, harder-to-get-to dock facility.

Port Canaveral is the largest industrial port on the Florida east coast. Recently the U.S. Coast Guard, which is part of the Department of Homeland Security, tried to close it and other deep-water ports to shrimpers who had moored there for years. "They've made it almost impossible for boats to tie up in the Cape Canaveral Barge Canal for the past three years," said Tiggitha Montgomery, a fierce attorney who represents shrimpers. At one meeting, she recalled, "it got so hot the Coast Guard said shrimpers were derelicts, drug smugglers, and criminals. Now the shrimp docks are in a compound that looks like a federal prison; they have to have a key to get in there. You have to have permission from the Homeland Security people to get into your own boat. The crew is not allowed off the boat, so it's hard to find crewmen. They can't walk into port."

The Coast Guard threatened one shrimp boat owner with an enormous daily fine when he moored during a hurricane. They had to keep it clear of traffic, they said, in case oil barges needed to come through. "We're fighting it, but if that stays in effect, that means there are no safe harbors on the central east coast of Florida, no place for fishermen to go in a hurricane. They have to go to Georgia."

A fisherman is always a stranger who lives on boats and keeps to himself. Some drink a lot or smoke pot. They wear ragged clothing, seldom join civic clubs, or even go to church, so they're considered lowlifes and make easy targets—much easier than the land-owning farmers whose pollution coming down the rivers from feed lots and fields is choking the Gulf of Mexico and polluting bays and esturaries in the South Atlantic with pig and chicken feces.

This was all vividly illustrated on a day when shrimpers came to the Florida capital in Tallahassee to protest a new state regulation that was being implemented on TEDs. Before the agenda item came up, a hundred of these rough working men and women sat in the audience with their callused hands and salt-worn clothing. But the governor and Cabinet ignored them as they passed out awards to citrus farmers for their bumper year of production. The outcast and castigated fishermen watched as the beaming farmers, one by one, were called up on the podium to be congratulated. After the governor and agriculture secretary made speeches on the tremendous economic contributions the growers had made to the state, they lowered their regulatory axe on the other group of working-class people who also devoted their lives to producing food—the shrimpers.

Perhaps the shrimpers were being shunned not just for conservation issues or because of sea turtles, but because the conflict between farmers and hunter-gatherers is thousands of years old. Early farmers had to fight off migrating hunters. When game was scarce, the raiders charged out of the woods, grabbed domestic animals, and stole the stores of food until the farmers could drive them off. Fishermen, who freely rove the seas, are to this day regarded with disdain by those who stay put, go to the office, and grow papers and money in the halls of government and bureaucracy.

Meanwhile, as humans squabble over who gets the resources, large areas of the seafloor are dying. When flounder, crabs, and shrimp came flopping, crawling, and jumping up on the beach in past years in Alabama's Mobile Bay, that was a Jubilee. It was caused by low oxygen levels in the water that happened during the summertime. Some of the oldest accounts of Jubilees date back to the 1860s. When it happened, people rushed down to the seashore and filled their coolers with fresh seafood.

"Jubilees—I grew up with Jubilees," said Jimbo Meador, the man who loosely inspired the novel and movie character Forrest Gump. While Forrest Gump was a simpleton, Jimbo Meador is extremely competent. "Me and an old black guy used to fish Mobile Bay together," he drawled. "We made our living catching fish, mainly flounder and blue crabs. The Jubilee was a natural condition of low dissolved oxygen in the water that came out of the river. When there was an east wind blowing, the sand eels started popping out of the ground. We ran to the beach with a cooler and started grabbing up fish before everyone else got there. Crabs get on top of the water, like they're trying to climb out, but it affected flounder the most, because they don't have swim bladders, and run up on shore to get away from the low-oxygen waters.

"The fish didn't die fifty years ago, when I was a boy. It only lasted a few hours, and when the wind changed it was over. They'd recover quick, and swim off. Don't confuse what's happening today with a Jubilee. A lot of what are called 'Jubilees' nowadays are toxic algae blooms that get into their gills and kill the fish," Jimbo said indignantly. "This dead zone is something much different—it's killing all our seafood. It's far larger. It doesn't last for hours; it goes on for half the year!

"And it's stupid to catch up all the menhaden that are filtering the plankton out of the water, turn them into fish meal to feed chickens, cows, and pigs so they can send their poop back down the river and contribute to the dead zone," he added angrily. "But as Forrest Gump said in the movie, 'Stupid is as stupid does!'"

Plankton, the microscopic floating plants and animals at the base of marine food chains, are beautiful. Drop a plankton net with a mesh as fine as a nylon stocking into rich estuarine waters at night, and it will mostly catch animals or zooplankton. Remove the jar attached to the end of the net and hold it up to the light, and it will swarm with copepods, larval stages of crabs, shrimp, and starfish, arrow worms, larval fish, and eggs. Most of the single-celled plants of the ocean are even smaller and pass through the tiny squares of the net. Use a net with a mesh finer than silk, and it will clog up with a greenish-yellow soup of phytoplankton that is the ocean's staff of life.

Dilute a drop of it with seawater and place it under a high-powered microscope, and a world of glittering jewels called diatoms will be revealed. In Greek, diatom means "cut in two" because one valve of their glass shell fits unevenly over the other half.

Of the approximately twenty thousand species of diatoms, some have fluted edges or appear as luminous green globes adorned with lace. If they were larger, their glass skeletons would be worn as necklaces, rings, and crown jewels. Diatom collectors around the world treasure these microscopic gems.

In some parts of the world, the glass skeletal remains of diatoms pile up into 700-foot layers and are mined as diatomaceous earth, which is used in swimming pool filters and for abrasives in toothpaste and car polish. Scientists also use the skeletons to track changes in the Earth's environment. Sediment core samples of Chesapeake Bay, for instance, show the impact of European settlers moving in and clearing the forests for farms. Silicate levels from diatom shells were constant for millennia until nutrient-enriched runoff into the bay caused major increases in diatoms in the water and silicate shells in the sediment starting in the 1700s.

Most diatoms avoid sinking out of the well-lit surface waters, floating due to a drop of oil in their cytoplasm that makes them lighter than seawater. Diatom oil from long-evaporated seas may be a major source of the world's petroleum. A single diatom can reproduce a hundred million descendents in a month. With proper nutrients and sunlight, a single gallon of seawater can grow a billion cells. There is now interest in farming diatoms to produce alternative energy.

They have a higher yield of burnable fuel than does corn and do not divert grain from the human food supply. Whether they float or sink, diatoms are the mainstay of aquaculture, since larval clams, oysters, salmon, and shrimp consume them with vigor. Penaeid shrimp have no trouble smashing through the diatoms' glass skeletons with their beaklike jaws.

Equally beautiful dinoflagellates are also found in a phytoplankton net. According to oceanographer C. P. Idyll, "The various species of dinoflagellates resemble Chinese hats, carnival masks, children's tops, urns, pots, and vessels of many kinds, the spiky knobs of medieval war clubs, balloons on strings, hand grenades, or lances."

Lacking tough shells to protect them, the dinoflagellates move under their own power using whiplike structures called flagella. Some of them sparkle with flashes of bioluminescence that make the sea look as if it is filled with glittering stars. Some use poisons to protect themselves from filter-feeding predators like clams. When conditions are right for those populations to explode, they become the source of toxic red tides.

Phytoplankton is nourished by the nitrogen and phosphorous coming down the rivers. A low level of nutrients, slowly released from the naturally decaying leaves and stems of marshland plants, is a good thing; it stimulates the water's productivity.

Oxygenless dead zones, however, are largely the result of burgeoning agricultural fertilizers washing off crops in the heartland of America. Farmers need phosphorous to grow their crops. So towering draglines, the size of multistory buildings, rip the skin off the earth and expose the phosphate-rich rocks near the Suwannee River and in other phosphate mines in Central Florida. Crushed into powder, it's shipped to the hungry croplands of the corn belt in Iowa, Indiana, and Illinois. And when it rains, nitrates and phosphates from fertilizers run off into the streams and rivers, flow down the Mississippi, and overfertilize coastal phytoplankton, which explodes. The microscopic plants die, fall to the bottom, and rot. Both the process of decomposition and the bacteria that bloom on the rotting material consume all the oxygen in the water. The overharvest of plankton-feeding fish like menhaden that used to keep plankton populations lower only makes the problem worse. A hypoxic, oxygenless zone devoid of marine life forms. When a dead zone gets too big, it sucks the life out of the sea until not a single living creature can be found on the bottom. When the oxygen-depleted water moves in, crabs come to the surface like they're trying to climb out, fish gasp, and shrimp disappear or die.

According to Dr. Kevin Craig of Florida State University, who has been studying the problem for years along with a team of scientists from Louisiana and Texas, natural hypoxia like Mobile Bay's Jubilee is rarely fatal and doesn't last long. "Hypoxia is a natural process to some extent, but we accelerated it, and we've been doing it for a hundred years. As farming operations increased, more nitrogen gets into the system, fueling the productivity of algae. We've seen a threefold increase since the 1950s, which is largely the result of agriculture and nitrogen-based fertilizers."

The mass of low-oxygen water that forms every summer off the Louisiana coast has roughly doubled in size during the last decade. Fisheries biologists and regulators were focused on the increasing number of shrimp trawlers and sea turtles in the '80s and '90s, but the burgeoning dead zone failed to catch their attention. The dead zone off the mouth of the Mississippi River covers 13,000 to 20,000 square kilometers (5,000 to 8,000 square miles), an area almost the size of Connecticut. It now threatens the half-billion-dollar-a-year Louisiana shrimp industry by creating an oxygenless barrier that disrupts the shrimp's migratory patterns. Some shrimp populations are forced out into deeper colder waters, where they grow slowly, and others are held at bay in the shallow warmer bays, where they overeat their food supply. Sometimes shrimpers drag around the edges of the dead zone fifteen miles offshore and catch shrimp, but often the big trawlers are forced inshore and drag the bays and compete with the smaller boats that catch the smaller shrimp.

The origins of the dead zone in the Gulf of Mexico are visible when you fly across the Midwest and look down on the vast tapestry of green and yellow squares of farm fields that cover the landscape. A few creeks and rivers are fringed with trees, and everything behind that is flat open fields. Often the fields run right down to the creek itself, leaving no wetland to buffer the fertilizers and pesticides from entering the water. Corn, wheat, soybeans, and sorghum grow in the imprints of long-drained watercourses.

Half of the fringe wetlands, potholes, and small lakes have been drained over the past hundred years, mostly for agriculture development. What the U.S. Army Corps of Engineers didn't change, the U.S. Soil and Water Conservation Agencies did. No creek or slough was too small for them to either fill or turn into a drainage ditch. At the time, they believed they were making the farmland more productive. But those cottonwood- and willow-fringed watercourses improve water quality, act as a sink for nutrients, filter out suspended solids, and absorb heavy metals.

To this day, however, keeping wetlands in a natural state draws strong opposition from some farmers. Once, while driving across Kansas, we stopped at a local cafe, where old men wearing coveralls had gathered for morning coffee. We listened to them complain bitterly about how their private property rights were being taken away because the government planned to require that they keep 10 percent of their land in wetlands just to save a few birds, by which they meant migratory waterfowl. The farmers cared nothing about the filtering ability of the wetlands, nor were they concerned with the fate of the Gulf of Mexico. Problems on the seafloor hundreds of miles away were simply meaningless to them.

Another dead zone off the coast of Oregon isn't related to runoff from fertilizer, but to deep-sea upwelling from 400 to 600 meters off the continental shelf. It didn't exist until 2002, but now it recurs each year and comes within 2,000 feet of the coast. Video of the bottom was collected by researchers at Oregon State University using remotely operated vehicles (ROVs) in 150-foot depths on the Oregon shelf. In 2001 they took pictures of giant plumose anemones, sea cucumbers, and black rock fish. They found over a hundred species, some of which can live a hundred years. The bottom was alive with spotted wolf eels, Pacific halibut, quillback rockfish, brittle stars, and sea stars.

Five years later, in 2006, they returned to the same spot after the most severe hypoxic event in Oregon's history. The ROV showed crab legs scattered on the bottom, dead anemones, bones and shells, dead worms, and no fish. Big white mats of bacteria covered the seafloor, which form only when no oxygen is present.

"Dead zones are not just limited to the U.S. or industrial countries. It's the same for developing countries, and it's growing worldwide," said Kevin Craig. "The hypoxic zone in the Gulf of Mexico is the third largest in the world. The largest, in the Baltic Sea, is about the size of the state of South Carolina.

"The Nile River in Egypt, the longest river in the world, has had a fourfold increase in phosphorous and a fivefold increase in nitrogen since the 1960s. The Yangtze River, the longest river in China, has a sixfold increase in nitrogen loading to the Yellow Sea. The fisheries have crashed in Shanghai. Today there are over four hundred dead zones around the world, and they have increased by one-third since 1995."

Kevin was studying croakers and spots, small fish that are common in the Gulf of Mexico bycatch, as well as brown shrimp. "On the Louisiana shelf when there's no hypoxia around, the shrimp are in about 60 feet of water. But when there's severe hypoxia, croakers get trapped inshore or get bumped offshore. You lose about half of the natural habitat for croakers, and shrimp lose about a quarter of their habitat. The water is colder offshore and there is less food, so they often grow slower.

"We know that shrimp typically avoid areas in which oxygen levels are less than two parts per million. It's a question of how long can they live in that low an oxygen level, and what the impacts of it are. It doesn't necessarily kill them. They are amazingly adaptable animals; they have mechanisms to cope with low oxygen. They can regulate their activity levels so they conserve oxygen, but those things incur costs. Shrimp at the edge have lower lipid levels than shrimp outside of it. That diverts energy from growth and increases physiological stress. We use aerial surveys to find the shrimp boats, and that tells us where the edges of the hypoxic zones are. Then we use our boats to map the low-oxygen areas and sample the shrimp and fish. Where you see shrimpers working, that's the edge of the dead zone."

Tim Thompson, a shrimper turned tugboat captain, recalled the days of working around dead zones when he worked on the shrimp boat *Stormy Seas* with his father.

"Back in the '70s we used to run between Cat Island and Chandeleur in Mississippi and dragged right through the dead zones. We never knew we were in one until we ran into it. Sometimes the water had a different color to it, kind of pale and hazy that's hard to describe. It had a bad odor to it, like a mudflat. This was something else; it stank like soured salt. You know how the salt air smells clean and fresh? I love that smell, but this was a sickly salty smell, kind of like rotting sea lettuce washing up on the beaches. When we got into those lifeless zones, there was nothing. We'd haul in the nets, and all we'd catch were dead, broken shells and dirt. Not even the sand dollars were alive.

"When we shook the junk out of the bag, water would splash on us, and it itched like hell. At first we thought it was some tentacles off of jellyfish, but it wasn't. There were no jelly, and it didn't sting; it just itched, irritated your skin, and made it break out into a rash. We had to use ammonia and meat tenderizer to get rid of it.

"The bottom looked different. Sometimes we'd snag one of the mud lumps with the try net, and it would bring up mud, but it sure didn't look like any other bottoms that we dragged. It was a kind of black mushy scum, thick like pudding, and was covered with an inch or two layer of white sand. It was nasty! It had a different feel. It looked dead and it felt dead, probably because all the worms in it were dead. We'd keep on running and dragging, catching nothing, and suddenly we'd pull up the try net, and there were shrimp in it. Sometimes it was pure shrimp, except for maybe a fish or two. I guess we found the edge."

Bobby Kirvin, who once owned the largest fish house in Apalachicola, worries about the future. "Shrimp are like fishes. If they have tails, they get gone. But they can only go so far. Shrimp sit around the edges of the dead zone and bunch up, like running a bunch of cows in a pasture, and they pile up at the fence. They get thick there because they can't go through. It's the same with shrimp. They leave the pasture when there's no oxygen and go to where they can breathe. People catch them, and the landings look normal, but they're not. The real problem is overpopulation. The farmers are producing more and more ethanol and fertilizer for corn production that runs off, making the dead zone grow larger."

Scientists are studying the problem, and politicians are considering ways to reduce the invisible nitrogen fertilizer runoff. But little actually is being done.

"In the Gulf we have an action plan to reduce the size of the hypoxic zone to 5,000 square kilometers. This year it's just under 20,000 square kilometers," said Kevin Craig. "There's a lot of effort to move farmers away from synthetic fertilizers, but most of it's voluntary, urging them to use no-till agriculture and replant with alfalfa to retain the nitrogen. There's some move to create buffers along the waterways, but at this point there's nothing required."

So, year after year the dead zone grows, fish decline, and shrimp grounds are lost. Each year the creeping dead zone grows like a cancer and poses a major threat to the central Gulf's multibillion-dollar-a-year seafood economy. We have no one to blame but ourselves. Conservation agencies regulate the size of the shrimp and the net, but they do almost nothing to stop land development and loss of wetlands. By regulating the shrimpers only, we are saving the paper clips while the corporation is going bankrupt.

Red tide is another natural phenomenon that has been made much worse by excess nutrients. Diatoms die off and dinoflagellates take over, many of which emit toxins. When the wrong species, such as *Karenina brevis* or *Pfisteria*, blooms, it creates a red or brown tide, which litters the beach with dead fish and makes the air reek with toxic fumes that cause people to cough and choke. Although the Florida Fish and Wildlife Conservation Commission doesn't want to admit it, circumstantial evidence is growing that links development to toxic algae blooms, sponge die-offs, and coral bleaching. With growing frequency, dead fish, dead dolphins, dead turtles, and dead birds wash up on Florida beaches. Coastal development and urban stormwater runoff are making red tide worse along the Florida Gulf Coast, as is nutrient-enriched water from the Mississippi River. Oddly enough, red tide has little effect on penaeid shrimp. Fish may be belly-up as far as the eye can see, and the water can stink, but the unsinkable shrimp are still crawling on the bottom, probably feeding on the fallen.

10

In Search of the Perfect Prawn

"Aquaculture is being developed to reduce the pressure on wild stocks," Al Stokes, director of the Waddell Mariculture Center in Beaufort, South Carolina, told us. "Since the shrimpers can't possibly keep up with the growing demand for shrimp, the production must come from the farm, where they can be grown in large quantities, delivering uniform-sized shrimp on demand. That will leave a niche market for wild shrimp. Fresh wild shrimp will be local, and ideally, as the demand goes up, there will be more boats, and they will catch it all up, and aquaculture markets will fill in the difference."

Dr. Motosaku Fujinaga was the father of modern shrimp farming. From 1930 to 1968, with an almost maniacal zeal, he learned how to spawn mother shrimp in captivity. He cultured their larvae, raised them to juveniles, and released them into ponds. He published paper after paper, corresponding with other scientists around the world, spreading his knowledge to anyone who would listen. Before his work, no one knew what shrimp larvae looked like, much less what to feed them.

The cheerful Japanese scientist who had a "wet thumb" gladly took money from the Japanese military government in World War II to research the life stages of penaeid shrimp and improve the nation's food supply. He discovered that females wouldn't produce eggs unless he fed them sea worms. He learned how to feed the larvae and raise them through their multiple growth stages. He discovered that larval shrimp had to be fed phytoplankton first and then be switched to zooplankton as they matured. He found that young shrimp grew best with dried squid pellets. He cultured plankton in the laboratory, trying to make commercial production possible. When the war was over he traveled to the United States in the 1950s, working with American scientists at the Bureau of Fisheries who were also experimenting with farming shrimp. Much of our knowledge of the growth and development of shrimp larvae came from this Johnny Appleseed of shrimp culture.

Aquaculture wasn't new; it had been practiced in Asia for thousands of years, but it was more ranching than farming. Small shrimp were seined from the estuaries, placed in tidal ponds and canals, fed with fish scraps, and allowed to grow up in a predator-free environment. The tides flowed in and out of the fences and nets, bringing clean water to imprisoned shrimp and taking away their waste. When the farmers wanted some to eat, they dragged nets in the ponds or drained them, and grabbed the jumping shrimp.

Fujinaga's dream was to start a shrimp farm in America. He surveyed both the South Atlantic and Gulf Coast of Florida with fisheries biologists, sampling the shrimp populations and collecting brood stock. He declared that Panama City, Florida was the best place to start a shrimp farm. Like Moses falling short of the Promised Land, Fujinaga personally did not succeed in doing it. But he sold the patent rights to an investment group of Japanese and American businessmen called Akima International, headed by a young man named Aki Kawaguchi, who was fluent in English. Kawaguchi approached the DuPont Corporation, trying to raise funds, but the giant chemical company had trouble conceiving of shrimp farming.

They had every reason to be wary. There was no shortage of unscrupulous pseudo-scientists to take advantage of the aquaculture boom at the time. More aquascam than aquaculture, a number of companies sprang up in Florida that had ponds seeded with shrimp and other species they claimed to have grown but that had been trawled from local estuaries. At the sight of several hundred pounds of jumping shrimp being seined out of small ponds and served at lavish dinners for would-be investors, money flowed.

Fortunately, serious scientists and businessmen also were trying to do it right. Ralston Purina jumped into developing artificial shrimp feeds and poured money into hatcheries in Crystal River and the Florida Keys. They experimented with raising pink and white shrimp in the warm-water discharge canals from the Crystal River nuclear power plant. This seemed like a good idea, except for the fact that the shrimp died from the antifoulant poisons added to keep the pipes from being clogged with barnacles. The King Ranch in Texas tried to raise shrimp, even though by then it was becoming clear that it was not profitable to grow shrimp that far north. It just wasn't tropical enough.

Hearing the scientists and investors of Akima International talk about the vast potential of shrimp farming, John Cheshire became infected with shrimp fever, quit his job at DuPont, and started Marifarms, Ltd. in the 1960s. With missionary zeal, he raised $17 million from his family and from friends at DuPont and hired Japanese scientists. Over the vociferous objections of environmentalists, he convinced the Florida state legislature to grant him a state submerged-land lease. Marifarms fenced off part of St. Andrews Bay in Panama City, built impoundments and ponds in the marshes, and started raising shrimp. The company's Japanese scientists built a state-of-the-art hatchery, hoping to breed a shrimp that could grow fast in overcrowded conditions.

From the beginning Marifarms in Panama City suffered biblical plagues of predators, diseases, and irate shrimpers. Angry at the loss of their fishing grounds, the local shrimpers tore down the fences, let the shrimp escape, and caught tons of them. Eventually Cheshire hired the shrimpers to work on the farm, and peace ensued. They chartered large shrimp boats to take their scientists out to trawl for gravid mother shrimp. Over several years of experimenting, they succeeded in getting white, pink, and brown shrimp to spawn in their laboratory. The eggs developed through the larval stages until they were big enough to release into the big outdoor ponds and raceways before being transferred to the holding pens in the bay.

For a decade the scientists and businessmen struggled with crabs eating holes in their nets, which leaked shrimp. They used a fish poison called rotenone in an attempt to control predatory fish, but pinfish were immune and feasted on juvenile shrimp. When low winter tides came, flocks of cormorants, pelicans, and seagulls dropped out of the sky and gorged themselves on Marifarms' shrimp dinners. Employees had to fire guns, blow horns, and rush out in airboats to chase off the birds.

"We didn't count on the winter tides and north winds that blew the water out of the bay and left the shrimp stranded on the mudflats," John told us one night at his home in Panama City. "The seagulls had a feast with our profits. Nor did we consider the tons of hydroids and sea squirts that covered our nets, which created a wall of resistance to waves and made it easy for the next hurricane to rip them apart. When a hurricane struck, it tore down our nets, overtopped the ponds, and released millions of shrimp, who swam happily out to sea. As soon as the storm passed, a fleet of shrimp boats rushed in, and we watched helplessly as the entire crop was hauled away."

The would-be ocean farmers rebuilt the fences and put more juvenile shrimp into their impoundments. They fed them by spraying a ground-up slurry of smelly fish from a barge that polluted the bottoms in their ponds. Disease swept through the ponds, killing most of the stock. With all the problems, a $50,000 loss was considered a good year.

Harvests were anemic at best, but it wasn't the birds, shrimpers, or fences that led to Marifarms' ultimate demise. The seasonal climate allowed only one crop a year. Shrimp farming would have to move to the tropics, where shrimp could grow year round, harvesting three crops instead of one. They tried to move to Ecuador. But Cheshire briefly landed in jail when he refused to pay off corrupt local officials, and unknown tropical viruses wiped out the shrimp. In the end, after two decades of struggle, Marifarms went bankrupt in 1980. Many of Cheshire's employees then moved on to other farms around the world, taking their technology with them (see Figures 10.1 and 10.2).

Figure 10.1 A shrimp feeding truck on a Texas shrimp farm

Courtesy of American Shrimp, LLC

Figure 10.2 Texas shrimp ponds with aerators running

Courtesy of American Shrimp, LLC

Meanwhile, Ralston Purina was trying to grow two other species of South American white shrimp in ponds in Brazil. With diseases and poor yields, these investors also had little to show for all their efforts, and they were running out of patience. It looked like shrimp farming was on the rocks. Desperately their scientists searched for a new shrimp that wouldn't expire in crammed pond conditions. They experimented with growing eight different species of penaeids until they finally settled on *Litopenaeus (=Penaeus) vannamei*.

The western white shrimp, *vannamei*, is native to the Pacific Coast of Central and South America and ranges from Mexico to Peru. Perhaps because it stays back in the mangrove swamps, it tolerates living in overcrowded conditions.

Vannamei reproduces best in murky water, likes algae, eats less meat, and doesn't compete with other shrimp; hence, it grows better. Investors were overjoyed when their eggs hatched and 110,000 post-larvae grew like crazy in their ponds, all without aeration. With a little care, and no predators to eat them, they calculated that when the shrimp grew up, they'd have the equivalent of four thousand pounds per acre. Again the sun shone for the pioneering shrimp farmers.

Years before farmers discovered *vannamei*, a zoologist named Willard Gibbs Vanname had collected the first specimen. The Yale professor was best known for his definitive monograph on sea squirts, his work with terrestrial and freshwater isopods, and his work in ornithology. In the obscure world of museum curators and carcinologists (those who study shrimp, crabs, and lobsters), history records that on March 25, 1926, Dr. Vanname purchased a male white shrimp in the fish markets of Panama City, Panama and pickled it for the American Museum of Natural History collection, where he was curator of marine invertebrates.

There it sat for five years, having turned red in the jar of alcohol, until a staff biologist at the museum, Miss Pearl Lee Boone, described it as a new species. Apparently she admired Dr. Vanname, so she named it *vannamei* after him. She declared it to be the analog of the North American white shrimp, *Litopenaeus (=Penaeus) setiferus*, that Linnaeus had described two centuries earlier. Her paper went on to detail the spine and eyestalks and measured its legs, pinchers, and male sexual organs. In the rules of taxonomy, the author who describes a species appears after the Latin name. With the recent name changes, today the shrimp is known as *Litopenaeus (=Penaeus) vannamei* Boone, 1933.

In any case, Pearl Lee Boone had a hard life. In those days women had few opportunities in science, were paid next to nothing, and seldom married or had families. They lived the lives of academic nuns. From the American Museum of Natural History Boone went to Miami and worked for an aquarium for a few years, described new species of crustaceans, and published papers on the morphology of spiny lobsters in scientific journals.

With the commercial rediscovery of *vannamei*, shrimp farm fever spread to South Carolina, where investors madly built ponds near salt marshes. Twenty-two shrimp farms were soon operating. But when Waddell Mariculture Center in Beaufort, South Carolina had an outbreak of white spot virus, people feared that it might spread to the native wild shrimp in the bays and sounds and destroy the shrimping industry. The worst fears of environmentalists were realized when many *vannamei*, possibly infected with the deadly white spot virus, somehow escaped into the wild. The state of South Carolina asked shrimpers to turn in everything they found mixed in with their catches of native white shrimp. On the radio, the shrimpers quickly changed the name. "We got some of them Vanna Whites," they'd announce, named after the ever-smiling hostess of television's *Wheel of Fortune*. Vanna Whites also escaped from ponds in Texas and in Mexico, but they never lasted long in the open sea, and white spot viruses did not cause mass mortalities there.

Meanwhile, at the Waddell Center the scientists drained and sterilized their tanks and started all over again with certified disease-free South American white shrimp, *Litopenaeus (=Penaeus) vannamei.* They also tried to develop disease-resistant local white shrimp on an experimental basis. Eventually all the South Carolina shrimp farms closed, unable to compete with cheap foreign farmed shrimp.

To squeeze out more profits and grow more shrimp to the acre, multinational corporations in tropical countries built their ponds too close together. Because the same brood stocks were shipped to hatcheries around the world, the diseases went with them. With overcrowding, both bacterial and viral diseases with names like hematopoietic necrosis virus and baculoviral midgut gland necrosis became rampant and pestilence spread from farm to farm. Some diseases attacked the larval shrimp; others ate into the shells of the adults. The deadliest, Tara Syndrome Virus, was carried in the guts and feces of aquatic insects and destroyed up to 90 percent of the stocks.

Epidemics spread across the globe from farm to farm via infected post-larvae or aquatic insects. In a few months, starting in 1993, the entire aquaculture shrimp industry crashed when it became infected with viral diseases. In 1994 white spot virus, brought over in a batch of South American *vannamei* brood stock shrimp, killed off most of China's profitable shrimp crop. Millions of pounds lay rotting on the bottom of the high-intensity "grow-out" ponds, turning the waters putrid, fouling air, and forcing China to turn away from high-intensity monoculture. In northern China at one point, the air was so foul from rotting shrimp farms that aboard airplanes, passengers gagged and held their breath on takeoff as the planes soared over the shrimp farms.

As we took off from the airport at Quingdao, China that same year en route to Shanghai, a business passenger turned to us. "I make this flight every week. You lucky. Two months ago, the shrimp farms get disease; the water turn black and stink so bad it take your breath away." Down below, the shoreline was carved into one vast network of shrimp farms that now lay empty.

Eventually many farm ponds became so disease-ridden and polluted that they had to be abandoned. When ponds were no longer viable, the shrimp farmers moved on, plowing up new mangroves like slash-and-burn farmers to make new ponds, destroying still more fisheries.

Scientists worked around the clock, desperately applying antibiotics that controlled bacterial but not viral diseases. Extremely powerful drugs such as nitrofuran and chloramphenical were mixed into shrimp feed even though they may cause liver disease in humans. Probiotics, beneficial bacteria similar to those found in yogurt, were added to the ponds to more rapidly digest and remove fecal and other waste from the mud. Larger farms spent millions adding bigger pumps and aerators to improve water circulation and reduce disease-breeding organic buildups on the bottoms of the ponds.

To combat viral diseases, farmers bred disease-resistant brood stock from the remaining handful of surviving shrimp and implemented strict quarantine procedures. Everything entering the farm was sterilized, and the shrimp were raised in operating-room-like sterile conditions. After several years and fifteen generations, domestic shrimp became disease-resistant.

After the diseases were dealt with, shrimp farms again proliferated from the Americas to Taiwan, China, Korea, Thailand, Indonesia, Vietnam, Malaysia, India, Bangladesh, South Africa, Madagascar, and other east African countries (see Figure 10.3). Giant multinational corporations such as BP, Ralston Purina, and A&P invested heavily and made a lot of money. The technicians and biologists who had worked for Marifarms were much in demand. They dispersed around the globe to tropical and semitropical countries where shrimp could be readily grown, setting up farms in Asia, Central and South America, Cuba, Africa, India, and Saudi Arabia. They experimented with local species because most countries were afraid of bringing in nonindigenous species like *vannamei* that might spread disease to their local shrimp.

Figure 10.3 An aerial view of a Madagascar shrimp farm

Photo courtesy of JM Amouroux, HydroTech Solutions

China's biologists, many trained at the National Marine Fisheries Service in Galveston's aquaculture labs, sliced and diced their coastline into a vast array of ponds and grew their native white shrimp, *Fenneropenaeus (=Penaeus) chinensis*. In the fishing town of Haiti in Shandong Province, the fisheries department took us to an experimental shrimp polyculture pond. A young man poled us into an impoundment, pulled up string overgrown with oysters, and shook a half-dozen tiger prawns and Chinese white shrimp into the skiff. He explained that the oysters, sea cucumbers, and tilapia helped filter out the excess phytoplankton and consumed bacteria that built up on the bottom of the ponds from uneaten shrimp food and waste, minimizing the chance of disease.

Japan kept on growing its shrimp, *Marsupenaeus (=Penaeus) japonica*, and farms in Africa, India, and the Middle East dug ponds and filled them with Indian Whites, *Fenneropenaeus (=Penaeus) indicus*. Brown and green tigers and banana shrimp were grown in Australia, and Thailand specialized in the big tiger prawns, *Penaeus monodon*. Costa Rica and other countries were raising freshwater prawns, *Macrobrachium*, the largest and perhaps least attractive-looking shrimp in the world. They attained the size of small lobsters, but the thick-shelled river shrimp were blue, didn't taste as good, and had less market.

All shrimp have individual differences. Some need diets higher in protein than others. Some prefer mud; some crawl about; some burrow in the sand or mud; others don't. Some like murky waters, others clear. Each species has advantages and disadvantages for farming. Some grow faster than others but might be more subject to diseases or intolerant of fluctuations in salinity. Of all the species of penaeids that shrimp farmers have attempted to grow worldwide, only the Pacific white-legged shrimp, *Litopenaeus (=Penaeus) vannamei*, and the black-striped tiger prawn, *Penaeus monodon*, have achieved stardom. By 2006, these two species made up 85 percent of all farmed shrimp.

Monodon is larger but has difficulty reproducing in captivity. Thailand is the world's largest producer of pond-raised glistening black-and-white-striped tiger prawns that flowed into American supermarkets, but that country has difficulty finding enough brood stock to meet the demand. Tiger prawn mothers are hard to come by. Hatcheries have paid $86 to $215 apiece for gravid tiger prawns. Brought from the ocean, like American pink and brown shrimp, they mate only when the females molt, their bodies are soft, and the males can penetrate them. After several years of successive generations being raised in ponds, the female population as a whole stops producing pheromones, and males are no longer attracted to them. Also, in places like India, semiclad men and women wade through muddy water in the mangroves, pulling small nets, hoping to find wild-caught post-larval tiger shrimp to raise.

World appetite for shrimp has grown so rapidly that many Thai river farmers converted their coastal rice fields into shrimp ponds. As much as five tons of shrimp a year can be produced from a pond the size of a football field, and rice farmers who had been making $4,500 a year suddenly saw profits of $20,000 to $40,000. Yet another case of shrimp fever broke out—a boom fueled by the World Bank and multinational corporations.

By the 1990s, an environmental backlash developed against the industry because it not only created epidemics but also destroyed wetlands, polluted adjacent water bodies with shrimp waste, and displaced rice farmers. When shrimp farms go bust, a wasteland of abandoned ponds, dead trees, and contaminated land and water is left behind. Just as the California gold rush destroyed the rivers and lands in the West, the shrimp gold rush devastated mangrove forests around the world. Thousands of acres of coastal mangrove forests were cleared to make room for acres of shrimp ponds. Thailand and Ecuador lost a quarter of their mangroves, and with them went the local fisheries. Eighty to ninety percent of commercial seafood species, such as crabs, fish, and shrimp, that inhabit the tropical oceans use mangroves as a nursery. Fishermen grew angry at the loss of wild seafood and their livelihoods. To make matters worse, fertilizer-enriched waters were discharged from the ponds, further degrading water quality and impacting seafood.

In Bangladesh thousands of farmers suffered when their rice crops were damaged by seepage of salt water from the shrimp farms. Investors bought up farmers' infertile lands and rendered the peasants landless. Three hundred thousand people were displaced in one province alone, causing violence to break out. When the tsunami struck the Indian Ocean in 2004, urban sprawl and shrimp farms in Bangladesh and Thailand were blamed for greater losses of human life, because the lost mangrove forests would have protected villages from much of the surge.

Ecuadorian and Malaysian environmental groups are calling for a boycott of cultured shrimp and a policy review of the entire shrimp industry. They do so with much justification, citing both the environmental destruction of trawler bycatch and the pollution and destructive practices of shrimp farms. But so far they haven't had much luck. Business is still booming, with 170,000 tons of farmed tiger prawn raised annually in Thailand alone. Aquaculture shrimp account for a quarter of world production, a worldwide market valued at $6 billion at the farm gate, and over $20 billion retail. Giant corporations like Aqua Star (BP), Wal-Mart, and Red Lobster keep buying more shrimp. Because of worldwide certification, all farmed shrimp must now be tested and certified as being uncontaminated with antibiotics and chemicals. Shrimp are being grown in ponds around the world like broiler chickens with fewer antibiotics. China has a bad reputation for using antibiotics, but Thailand claims that because it has spent millions improving the ponds and putting in liners, it uses no antibiotics.

Even the shrimp farmers agree that criticism has been good for the industry. It has caused them to upgrade and improve health and environmental standards—not only for the benefit of the consumers, but also for their workers. Although environmental and human-rights groups still protest, better environmental standards have been developed. The fleets of trawlers have remained much the same. But shrimp farming around the world has undergone rapid evolution over the past twenty years—not just in improving the health of the shrimp stocks, but by minimizing the industry's impact on the environment and human rights. A Thai farm described in *Time* magazine as keeping its workers in slavery, in brutal conditions behind barbed-wire fences and working them round the clock, is no longer in business. The World Wildlife Fund and other organizations are pressuring the industry to develop and implement better standards—not just on paper, but on the ground.

In the past, farms reported low daily nitrogen levels being discharged into receiving bodies of water as they circulated seawater through their ponds. But then they failed to report the discharge of highly polluted water when the ponds were harvested. The shrimp would churn up the bottom mud, which was loaded with shrimp waste, uneaten food, and antibiotics. This effluent can kill larval and juvenile fish, shrimp, and crabs entering the estuaries. Under the new standards, participating farms promise to stop discharging harvest waters and will reduce the amount of wild-caught anchovies in formulated feeds to keep the oceans from being depleted. No loss of mangroves will be permitted. If such standards are implemented worldwide, consumers can be assured that they're eating shrimp that have much less impact on the environment than in the past. Wal-Mart, the word's largest retailer, just adopted a policy of purchasing only farm-raised shrimp from companies that are certified by the Global Aquaculture Alliance, which sets rigorous standards to protect the environment and produce healthier products. Slowly but surely, inefficient and badly built farms have gone out of business. They are being replaced by farms that are bending to global pressure, growing antibiotic-free shrimp, and no longer destroying mangrove swamps. Some companies are even planting new mangrove forests and attempting to restore the environment.

Although Asia and Madagascar prefer growing tiger prawns, *vannamei* is rapidly beating out the various local species because they are more disease resistant, grow faster, are less finicky feeders, and can better tolerate living in crowded ponds. Furthermore, it's easy to get *vannamei* brood stock. Today, hatcheries maintain captive stocks of pathogen-free *vannamei*, some of which have lineages going back thirty years. In a few short years *vannamei* became the main production species and is the leading farm-raised species in the Western Hemisphere, representing more than 99 percent of production.

Now, courtesy of worldwide aquaculture, certifications, and truth in seafood packaging campaigns, the name *Litopenaeus (=Penaeus) vannamei* is sometimes printed on cartons of imported farmed shrimp. Also, the name Vannamei occasionally appears on little plastic signs sticking up from beds of ice and shrimp in seafood cases in supermarkets. John Cheshire, the founder of Marifarms and a leader in world aquaculture, would declare the species to be "the most valuable invertebrate in the world." When you see Vannamei on bags of shrimp in the supermarket, send kind thoughts to the memory of Pearl Lee Boone, who first described it.

Shrimp are being successfully raised in over 300,000 shrimp farms today. Aquaculture shrimp now account for almost half of the world's production, with a worldwide market valued at $7 billion to the farm and over $20 billion in retail sales. Only 10 percent of those succulent crustaceans with their delicious pink-and-white flesh consumed in the United States come from domestic shrimp boats. The rest are grown on foreign shrimp farms or caught on foreign boats.

Sixty percent of the 4.1 pounds of shrimp that the average American eats today are *Litopenaeus (=Penaeus) vannamei* and the tiger prawn, *Penaeus monodon*, which are pond-raised in Asia, Africa, and the Middle East. Cheshire and his original investors lost everything, but he remained proud of the fact that Marifarms' efforts resulted in today's $20 billion worldwide shrimp-farming industry.

Shrimp farms produce no great adventures. There are no storms at sea or boats going down in waters seething with sharks. In the book, Forrest Gump was a shrimp farmer, but in the movie he was a "shrimping-boat captain." In the industry publication *Shrimp News*, author Winston Groom explained why. "Shrimp farming is like watching grass grow, whereas if you've got a shrimp boat, you've got motion and action, and that's what movies are all about. That's why they call them movies."

Rod McNeil, a biologist who spent a lifetime growing shrimp around the world, found a way to pass the time. He spent many hours on the bottom of shrimp ponds to see what the crustaceans were up to and described it to Bob Rosenberry in *Shrimp News*:

> *"Suddenly as I lie motionless on the bottom of the pond, a troop of thousands of shrimp approaches... looking for food. They move at a pretty good clip, with the very largest shrimp at the head of the troop and the smallest shrimp at the back, or not really with the troop at all. They move together somewhat the way a school of fish moves, in unison. When threatened by a predator, they scatter the same way a school of fish scatters, confusing the predators and making it difficult for them to zero in on individual animals.*

"The troop is shaped like a teardrop, with the bulbous portion forming the leading edge, and the biggest shrimp across its front, then tapering off to the smaller shrimp at the rear.... It's kind of scary when you're lying on the bottom and watching them come toward you. Lying motionless and silent on the bottom, I watch their approach, and their reaction to me. When they spot me, all their antennae go up. I move toward them; the big shrimp on the leading edge of the troop take a head-high posture and keep an eye on me. I don't move in any closer than a meter and a half. They freeze, watch me for thirty or forty seconds, and then troop right over the top of me. I dive with a wetsuit, but my hands and head are exposed, and they nibble on the skin, pinch it here and there, trying to figure out how to eat it.

"There's a very strong preference for natural foods, particularly at night, when shrimp become solo hunters. During the day, when they're trooping around the pond, their antennae are swept back along their sides. When the antennae lock on to a food source, they point directly at it and sometimes twitch. They can sense the extremely small electrical fields generated by the animals that live in the pond bottom. When they locate something, they move their antennae over the spot and then start digging. It's tough to see what they pull out of the bottom because they tear it apart so quickly. They're vicious little critters.

"When they arrive (at the feeding tray) it's absolute mayhem. I'm typically two or three feet away, lying flat on the bottom, and it's actually scary. The big shrimp at the head of the troop get there first and swarm over the feed. As the smaller shrimp arrive,

they clamber on top of the big shrimp and before you know it, the tray is covered with five or six layers of shrimp. The big ones on the bottom attempt to throw anything smaller than themselves off the tray. A shrimp's life is always on the move, always eating."

Meanwhile, American shrimpers urge consumers to boycott farmed shrimp and eat the wild-caught product. When the venerable Senator Robert Byrd of West Virginia modified the Tariff Act of 1930 and pushed his Continued Dumping and Subsidy Offset Act (CDSOA) through Congress to help the ailing steel industry, he inadvertently issued a license to steal. Attorneys and lobbyists for the shrimp industry claimed that shrimp-farming nations were unfairly subsidizing shrimp farmers by building hatcheries and farms so that they could sell their products at below market prices in the United States. Before his amendment was repealed, Senator Byrd made it possible for American shrimpers to collect big cash rewards from offending countries by claiming they were economically injured.

This created an $800 million gold rush over a two-year period, reminiscent of the California gold rush of 1847 and the bank bailouts of 2008. The hastily formed Southern Shrimp Alliance petitioned the U.S. Department of Commerce to file suit in the World Trade Organization court against six major farmed-shrimp-exporting countries: Brazil, China, Ecuador, India, Thailand, and Vietnam. The Southern Shrimp Alliance told the World Trade Organization, "There are too many people's livelihood at risk and too many communities in the Gulf Coast and Southeast Atlantic facing ruin to ignore the devastation caused by unfair trade. This industry cannot be sacrificed for empty promises that our communities will, at some indeterminable point in the future, enjoy the benefits of free trade."

Exporting countries argued that it wasn't a matter of illegal subsidies. They were selling shrimp cheaper because they were more efficient, had lower labor and fuel costs than trawlers, and produced a better, more reliable product that was much in demand. The tropical shrimp farms, they said, enabled foreign producers to harvest more shrimp than the ocean provided and to do so more efficiently. Hence, they could sell their shrimp at a lower price. If import tariffs were put in place, the price of shrimp could increase by 44 percent in the U.S.

The World Trade Organization agreed that the shrimpers could claim damages. The Southern Shrimp Alliance was awarded duties that ranged from 7.6 percent from Ecuador to 112 percent from China and India. U.S. Customs would collect the money from these governments and give it directly to the offended industry. The Southern Shrimp Association said that the additions were merely "a step in the right direction" and asked for additional tariffs of up to 200 percent.

The money that U.S. Customs collected was to be used to help the shrimpers and their infrastructure to keep them from going out of business. Most of the personnel of the U.S. Customs Service had never seen a shrimp trawler and suddenly had to decide who was eligible to receive the funds. How does one pay the owners of over ten thousand shrimp boats that were scattered from North Carolina to Texas?

They didn't. Instead, they gave $102 million worth of import tariffs that they collected from Brazil, China, Ecuador, India, Thailand, and Vietnam in 2005 to lobbyists, attorneys, and industry reps of the Southern Shrimp Alliance who brought trade actions and filed lawsuits. You had to be a member of the organization to get on the payout list. Most of the shrimpers got only a pittance, an average of $8,000 to $20,000. The money was tied to their production of the previous year. If a shrimper had a bad season, his boat broke down, or the weather was too rough to work, he didn't get much.

Jim Lycett, a Carrabelle shrimper who was a member of the SSA, complained, "There was no oversight by the government. The money went to the fish houses, who weren't impacted like we were. I have three boats, and in two and a half years, with well-documented receipts, all I got was $15,000—chump change! The big corporations that own hundreds of boats grabbed most of the money." Everybody sued everybody, and claims of fraud were more common than shrimp on the seafloor.

11

The Miracles of Chitin

The future of shrimp farming and wild harvest both dance on a knife's edge that could go either way. Shrimp boats and their gear can be redesigned to cut down on bycatch and save turtles. Skimmer trawls are coming into use that don't tear up the bottom, and lighter nets that are less destructive are also more fuel-efficient. Shrimp farms are also upgrading, getting farther away from the polluted mud holes that festered with diseases and destroyed the mangrove swamps. Larger, better-financed organizations are using lined ponds with low risk to the environment and high-tech greenhouses to increase their yields.

A new vision is emerging—not just for the big money, but for poor people as well. Consider the Seawater Farm that briefly existed in the war-torn African country of Eritrea on the Red Sea. A group of visionaries from Arizona set out to build a totally self-contained shrimp farm by laying miles of pipe over the desert from the sea into 230 brick-and-concrete tanks. Working with impoverished villagers on the brink of starvation, they built a hatchery, raised shrimp in three artificial salt lakes, and grew shrimp and tilapia for export. The wastewater, rich in shrimp waste, flowed into fields of sea asparagus, a succulent *Salicornia* plant that one sees growing at the edge of the sea almost everywhere.

The thin crunchy green plant made good feed for goats and cattle, but it could also be eaten as a vegetable. Its protein-rich oils were extracted from its seeds, and when dried and crushed, its stems were turned into fuel briquettes or particleboard. The water then flowed into a wetland forest of 200,000 mangrove trees planted by villagers. This man-made wetland attracted more than 154 species of birds—123 of them on the endangered list—before the farm water returned, filtered, to the sea.

As Ned Daugherty, a university researcher on the project, said, "We believe firmly that projects like this can bring wealth and peace if done in a good fashion. We want to enhance the environment, generate wealth, and bring beauty and industry to the Eritrean people."

Unfortunately, they never got the chance. For a while prosperity reigned, and villagers raised goats and cattle and made money selling shrimp and fish. But in the end war closed in, the place was bombed, and they were all scattered. But the concept lives on and can again be a reality in the future. One farm in Ecuador is now producing organic shrimp, free of antibiotics and chemicals, and spraying wastewater on crop lands and growing jungle gardens filled with parrots and wildlife Another farm in Costa Rica processes their shrimp shells into chitin.

Whether farmed or wild-caught, the shells of the shrimp may someday be more valuable than the shrimp themselves. Chitin and its derivative, chitosan, are being used to make valuable products for the medical, food, fiber, and textile industries. Chitin is used for everything from medical plasters and bandages to fat blockers and industrial water purification agents. Chitin is a polysaccharide, made of sugars and starches. It makes textiles more washable and is used in medical braces and supports. If only there were enough shrimp shells to go around!

In the fall of the year, just off the beach of Alligator Point in the Florida panhandle, white shrimp forage on the bottom, tasting and eating everything. They forage on the living and scavenge on the dead, growing and molting as they go. Rarely does one find their molted chitinous shells on the beach. Those are quickly eaten by other animals or broken down by bacteria and the scour of sand. If it weren't for chitin-eating bacteria, the ocean would be clogged with the shells of crustaceans—not just from the crabs, lobsters, and shrimp, but from the trillions upon trillions of copepods and other minute crustaceans that are constantly shedding their shells. Krill, the favorite food of whales, change their skins nine times a year.

All those life-forms can be sorted into three major chemical categories. Seashells and sand dollars are largely made of calcium carbonate. The bodies of soft-bodied or spongy creatures like jellyfish, sea whips, and sponges are largely composed of collagen. The crustaceans and bryozoans are made of chitin, with a mix of calcium. The word "chitin" comes from

Greek, meaning "coat of mail." After cellulose, chitin is the most abundant organic fiber on Earth. It was first extracted from mushrooms in 1811.

A shrimp molts by first laying down a sticky foundation of this starchlike polymer. Then it pulls calcium out of the seawater to build a strong lattice to armor itself against the jaws and claws of the sea. When shrimp molt, they take up water until their carapace splits apart and they shake their way out of their "tails." As the carapace comes off, the shrimp contracts and pops out of its shell. They extract their appendages from the old shell much like you pull your arms out of the sleeves of a coat. Then they shake themselves out of their tails. When they finish, they are so exhausted they can scarcely move. So they seek a hiding place to get away from their predators until their shells harden, which takes a day or two. A shrimp molts throughout its lifetime, which is only a year or so. Like other crustaceans, when they grow old, they no longer molt, their shells become brittle, and they grow feeble.

Molting is a crazy way to grow—a difficult and exhausting time for any arthropod. The jointed-legged must grow from the inside, and when they get too big, they pop off their shells. Crabs and shrimp crawl backwards out of their shells. Only horseshoe crabs walk forward, leaving a perfect replica of themselves behind which beach combers often mistake for dead crabs. The molts are no more or less dead than your human hair lying on the barbershop floor.

Before the 1970s most shoreside processing plants shoveled the shrimp heads and shells off the docks and into the river, to the delight of waiting catfish and crabs. For example, of the twenty thousand tons of blue crab harvested from

North Carolina, eighteen thousand tons of it was waste that made rural coastal towns smell like fish markets. As demand for peeled and deveined shrimp grew, the shells of shrimp and crabs piled up. With the rise of real estate and upscale condominiums being built on the coast, developers and newcomers objected to the smell. So the EPA wrote regulations prohibiting the dumping of untreated shellfish waste in coastal waters. It could have been used for soup stock, but it had to be buried in sanitary landfills. That created other problems. The seafood garbage took up too much space, and it rotted and stank worse than it did before. Finding new uses for crustacean waste became more urgent. Government and university laboratories experimented with grinding it into fertilizer and compost, but this was not cost-effective.

Many fish house owners were enthused about the idea of producing chitin, but commercial processing of chitin isn't easy. First the meatless remains are pulverized and then soaked in a bacteria broth to strip protein from the shells. Next it's boiled in a strong alkaline solution like sodium hydroxide—the same stuff used in Drano to clean out stopped-up sinks. It is then immersed in hydrochloric acid to neutralize the base. If you take a crab and boil away the calcium part of the shell with hydrochloric acid, you're left with a leathery covering made of chitin, a mucopolysaccharide that has the amazing ability to bind calcium. Finally, the shells end up as odorless dried flakes and powder that tastes nothing like shrimp and is ready for the bulk market.

After seeing how much handling and labor were required to process the shells, Americans went back to dumping them in landfills or in rivers when they could get away with it. What should have been an economic boost to fish houses along the South and Gulf of Mexico coast wasn't. It took too much handling for American businessmen, and there wasn't enough money in it for them. But for China and the Middle East, it was a bonanza. Suddenly, with chitin and its derivatives available, science and industry explored its use, and medical, agricultural, and industrial markets grew. Chitin, broken into chitosan and then into glucosamine, turned into a multibillion-dollar business.

Most of the chitin produced in the world comes from China, the biggest shrimp producer in the world. Iceland comes in second, producing tiny 100-count cold-water "salad shrimp" caught on factory-ship trawlers. Even with all the thousands of shrimp farms, getting enough shrimp shells on a regular basis is hard, so prices of raw materials have been rapidly going up. Although the bulk of chitin on the market comes from shrimp farms, the shrimpers still have a good opportunity, because the shells of wild shrimp are thicker, are of better quality, and have more chitin than farmed shrimp. Many farms are inland and successfully raise vannamei shrimp in well water, but the low salinity makes their shells soft, thin, and a bit slimy. Super-salty water, which comes from groundwater, isn't good either; the shells become hard and brittle. Unless the biomedical industry starts mass-producing mushrooms, other fungi, and insects, chitin products will remain in short supply.

Manufacturers buy the raw product and dissolve the chitosan flakes in special solvents to produce more valued derivatives like glucosamine, which is used to treat arthritis. Chondroitin sulfate and glucosamine are now sold in drugstores everywhere to relieve arthritis pain and restore movement. Glucosamine has been clinically proven to slow the progression of osteoarthritis in humans and help with hip dysplasia in pets. It is a key component of healthy cartilage. However, as we grow older, we lose the ability to produce enough of it, and the synovial fluid that lubricates our joints is reduced. As a result, we grow stiff, and movement is painful. Taking glucosamine regularly helps restore joint function. It usually takes a few weeks before we're ready to give thanks to a crab.

It's been known for years that chitin has wound-healing properties. For centuries Koreans have ground up the internal shells or pens of squids to treat abrasions, and Mexicans pack mushrooms into lacerations to accelerate wound healing. By the mid-1950s, chitin-coated sutures that enhanced healing were being used. Researchers at the University of Delaware Sea Grant College Program figured out a way to spin pure chitin filaments into sutures that could be absorbed by the body instead of being surgically removed. It's no longer necessary to remove chitin-derived wound dressings, bandages, and plasters from surface wounds, skin grafts, and burns. The reapplication of dressings is especially painful when they are attached to healing skin. The U.S. military has invested nearly $30 million because it proved under combat conditions in Iraq that chitin-impregnated bandages have been able to stop even profuse bleeding from bullet wounds in less than a minute and save lives. Processed shrimp shells of *Pandalus borealis* from Iceland are now being used as wound dressing.

According to one company that manufactures nasal sprays out of ground-up shrimp shells, chitin is a magic bullet when it comes to alleviating several allergy and sinus miseries. It's even being used for birth control, as an ingredient in a new sperm-paralyzing contraceptive. Sold as gels, liquids, films, pills, capsules, and powders, the worldwide market for chitin products is now in the billions of dollars. Rexall Pharmaceutical reported sales of its anti-arthritis formula in excess of $40 million in one year.

Reading about the wound-healing properties of chitin is one thing, but experiencing them is another. One day, I (Jack) managed to cut myself good on a piece of sharp plastic. As my slit finger gushed blood, I bound it tightly with paper towels, trying to stop the bleeding, but I ended up with a bloody mess. By the third change, I remembered what I'd been thinking about for the past couple of days—chitin. I rummaged around and found a bottle of "Natural Fat Trapper" pills made from chitosan. Still bleeding, I opened the bottle, shook out a capsule, and emptied the brownish-white flakes into the deep cut on my finger to see what would happen. Presto—it stopped bleeding instantly. It happened so fast that it was shocking. For the first time ever, the dance of wrapping one's hand, changing bandages, and waiting for the blood to clot was over.

The next day I washed it off, and to my amazement, the wound was all but healed. Planning to keep bottles of these capsules in our first-aid kits, I went to the drugstore and found that they no longer carried these pills. The manufacturer had been hit with a $10 million fine from the U.S. Food and Drug Administration for claiming the product caused people to lose weight without scientific data and clinical trials to back it up. Scientists are skeptical that it can be used as a "fat blocker," even though the Internet is filled with weight-loss testimonials and claims that say, "If you take chitosan, you'll slim down even if you eat a lot." Eat a big steak, pop a chitin pill to bind the fat, and the purveyors say it will pass through your digestive system without accumulating on your rump, belly, or thighs.

LipoSan Ultra (the Icelandic company whose trademark appeared on the bottle) reemerged with a different name, this time providing more evidence that the product did block the absorption of fat. It allegedly helps treat irritable bowel syndrome, constipation, and other digestive disorders and can make cheese and milk products more digestible. It's also employed in the inhibition of tumors and dental plaque and for blood cholesterol control.

Surgeons are now using it for antibacterial sponges for deep wounds and for hospital dressings. Chitin plasters are being tested at medical facilities for burn treatments. Not only does it protect and soothe the injury, but its natural antibacterial and antifungal properties also prevent the spread of infection. Contact lenses are now being made with chitin because it's tough, transparent, water-absorbent, and highly moldable. It finds its way into household products like diapers, sponges, feminine napkins and tampons, makeup powder, nail polish, and body creams. You may even be brushing your teeth with it.

Because chitin also grabs heavy metals and precipitates them, it is used in water treatment plants, and even to stabilize the muddy runoff from construction sites. Chitin removes industrial and toxic waste before it goes into lakes, rivers, and streams. Some companies prefer it for water purification because it binds ions better than the more widely used cellulose products.

Seed producers are spraying it on wheat, barley, oats, and peas to control nematodes and fungi. Exactly what mechanisms it uses to suppress fungus in wheat, barley, and oats is unknown and is still being investigated. Chitosan solution also increases the shelf life of eggs and is getting wide use in the food industry. Because chitin is nontoxic and biodegradable, it makes an excellent film for coating vegetables and other food products.

Scientists are experimenting with chitin as a possible replacement for plastics—the nightmare that never goes away. Plastic bags break down into tiny indigestible flakes and particles that kill marine life. It's been estimated that 30 percent of ocean fish have plastic particles in their stomachs, which interferes with their digestion. You can even find microparticles of plastic bags in the digestive tracts of sea cucumbers out on the tide flats. In some parts of the ocean, the floating particles are denser than plankton. Imagine a degradable chitin-based fishing line or drift nets that would disappear before they could entangle seabirds and turtles. Materials scientists are using chitin to create materials that can self-repair tears when exposed to ultraviolet light. It may someday result in paints and countertops that self-repair scratches, as well as many other industrial applications.

The University of Maryland is creating a chitosan-based nanoscale sensor system that can detect minute quantities of explosives and dangerous chemicals in the air and water to improve security in airports, hospitals, and other public facilities.

Alternative-medicine and health-food buffs use the derivatives of crab and shrimp shell for detoxification therapy, claiming that it cleanses the body of harmful metals and pollutants when combined with fasting and colon therapy. Japanese businessmen fork over a hundred dollars for a bottle of concentrated king crab chitin as a cure for hangovers.

There are even indications that chitosan could become part of a treatment for cancer. According to the National Cancer Institute, nanoparticles of chitosan, which are widely used to carry antitumor drugs and anticancer genes, are toxic to certain types of malignant cells. Studies in China demonstrated that chitosan nanoparticles, when delivered orally, may destroy human liver cancer cells growing in mice.

Not only is chitin used to filter swimming pools and spas, but it also cleans the water of nuclear power plant reactor tanks. Farmers love it because it boosts their plants' immune systems, repels insects, and increases yields. Coastal residents say they get spectacular tomatoes and marijuana plants simply by tilling crab garbage from the picking houses into the soil.

We need a new vision. Perhaps one day the pollution will be stopped and the oceans managed in a way that preserves marine life and benefits humankind. Why not ranch penaeid shrimp, which grow like insects and take only a year to mature? With healthy seas, they could be one of the most productive food sources in the world. If hurricanes can boost productivity by infusing nutrients, why can't we put shrimp into the sea? Perhaps shrimp could be conditioned to respond to a bell while they're being raised in hatcheries. After they're turned loose, they could be rounded up by sea shepherds at harvest time.

China, Japan, Australia, and Cuba are already using hatchery technology to grow larvae and then are releasing them into the sea so that fishermen can catch them. No one can say for sure whether all the millions of baby shrimp make a difference or just fatten fish. In this country skeptics insist that putting shrimp in the sea is like carrying coals to Newcastle. Would the shrimp born in hatchery tanks be suitable to release into the sea and still survive? Would the multibillion-dollar recreational fishery world improve? It's not a new concept. In Cuba and China, biologists claim that their commercial yields have increased by 50 percent since they began releasing hatchery-raised juvenile shrimp. The U.S. government has been raising and releasing striped bass and red drum for generations.

If the oceans are dying, as so many marine scientists are reporting, a more long-range vision is required. We need a vision in which the tragedy of the commons gives way to recognition that the future of every human being is linked to the well-being of every other human being. This depends in turn on a healthy environment and sustainable science-based resource use. Some would claim that the global cooperation necessary to achieve that goal is beyond the reach of human nature. But looming global-scale environmental crises will force us all to once again confront what Benjamin Franklin said in another time of great challenge: "We must all hang together, or assuredly we shall all hang separately." A time of great crisis is also a time of great opportunity—a chance to make big changes that cannot be accomplished in more ordinary times.

Acknowledgments

There are many people who we would like to thank for their help in providing information and editing the manuscript. Some of those who willingly gave their time for interview are Pam and Jim Lycett; Johnnie Keith, Vance Millender, Bobby Kirvin and many others advised us on shrimping. Manuel Espijo and Rick Eager, Tom Harper and Bob Rosenberry and the late Scott Quackenbush helped us with aquaculture. Without Holly Blair shepherding me (Jack) through South Carolina, making contacts and arranging for meeting, the work would have been far more difficult. Dr. William F. Herrnkind, Dr. Gregory C. Jensen and Winston Groom helped edit the manuscript, and encouraged us to write it, and Ben Higgins provided insights on turtle conservation issues as well as an artistic drawing. There were others who helped also and we thank them.

About the Authors

Jack and Anne Rudloe operate Gulf Specimen Marine Laboratories, Inc.—a unique environmental education center and public aquarium that supplies marine organisms to schools and research laboratories—in Panacea, Florida. (*See their web site at www.gulfspecimen.org.*) Jack has spent years working on shrimp boats, participated in the International Indian Ocean Expedition and has made collecting trips for the New York Aquarium and the National Cancer Institute. He has written six highly acclaimed and widely reviewed nonfiction books on the sea. His novel *Potluck,* a story of shrimping and smuggling, has been called a Florida classic.

Anne Rudloe has published two nonfiction books, *Priceless Florida* and *Butterflies on a Sea Wind*—a book on Zen mediation—and has also written many scientific research papers. Together, Anne and her husband have published articles in *National Geographic, Smithsonian, Sports Illustrated, Natural History, Audubon,* and other magazines. They have been featured in several major television documentaries on PBS and Jack has been interviewed on *The Today Show*, *Good Morning America*, Fox Network, and National Public Radio. In addition to writing columns for the *Tallahassee Democrat*, Anne is also an adjunct professor at Florida State University.

INDEX

D

E

Q

R

S

T

U

V

W-X-Y-Z

FT Press Science

Breakthrough!: How the 10 Greatest Discoveries in Medicine Saved Millions and Changed Our View of the World

Jon Queijo

0137137486 / 9780137137480
Available April 2010

Read the following excerpt from *Breakthrough!...*

10 World-Changing Revolutions in Medicine... and the Remarkable Human Discoveries That Made Them Possible

- The unforgettable life-or-death stories behind antibiotics, vaccines, DNA, X-rays, and more
- What happened, how it happened, and what it means to you *today*
- For *everyone* interested in science or medicine

Why are you alive *right now*? Chances are, you owe your life to one of the remarkable medical discoveries in this book.

Maybe it was vaccines. Or antibiotics. Or X-rays. Revolutionary medical breakthroughs like these haven't just changed the way we treat disease: They've transformed how we understand ourselves and the world we live in.

In ***Breakthrough!***, Jon Queijo tells the hidden stories behind 10 of history's most amazing medical discoveries. This isn't dry history: These are life-and-death mysteries uncovered, tales of passionate, often-mocked individuals who stood their ground and were proven right. From germs to genetics, the ancient Hippocrates to the cutting edge, these are stories that have changed the world—and, quite likely, *saved your life*.

The World's First Physician: HIPPOCRATES AND THE DISCOVERY OF MEDICINE 1

The Greek island of Kos, located in the crystal clear waters of the Aegean Sea and bordered by 70 miles of golden beaches, might be one of the best spots on the planet to fall ill—or to simply remain well.

Part of a 12-island archipelago, Kos is 200 miles southeast of Athens and just a few miles off the southwest coast of Turkey. Long, narrow, and verdant with lush foliage, the island is flat except for two low mountains along its southern coast. But it is in the town of Kos, an ancient village on the northeastern coast of the island, where the magic and medicine of this island begin.

It is tempting to speculate that the legendary history of Kos arises from its nurturing rich soil and abundant groundwater: Visitors entering the village are greeted by a lush landscape of tall palms, cypresses, pine trees, jasmines and, for an added splash of color, the bright reds, pinks, and oranges of the hibiscus. But if you want to locate the true pulse of Kos and its 2,500-year-old legacy, you must continue your journey…

First, face west and walk two and a half miles out of the village. There, amidst more lush landscape, you will approach a sloping site. Hiking up this slope, you pass an extensive complex of ancient ruins that rise around you in a series of terraces. Put aside your curiosity and continue to climb. Before long, you will arrive at a pinnacle. Gazing out from this high point, you stop flat in your tracks: The world has split apart.

Spread out before you is a breathtaking view of the Aegean seacoast. Inhaling the fresh coastal air, you feel the stirrings of the true spirit of this small island, the mystery of where two worlds meet. One, the "inner" world, is simply you: the tightly wrapped sac of blood and bone, emotion and mind, which is your body. The other "outer" world is merely every other thing in the physical universe surrounding you.

If you ponder for a moment the possibility *that two such worlds not only exist, but co-exist in a place we may not yet fully understand, then congratulations. You have finally begun to arrive, physically and metaphysically, on the island of Kos. For this is the place where, in the view of the world's first "rational physician," all life, death, health, and disease—and hence the practice of medicine and healing itself—begins.*

* * *

This ancient site is known as the *Asklepieion*, the generic Greek word for any "healing temple." But the Asklepieion of Kos is a temple like no other. Although today a crumbling ruin of broken walls, roofless chambers, and lonely columns supporting only air, in its heyday this was a bustling center of healing. Here patients in all stages of sickness and injury sought the best treatment they could find.

If you were suffering from disease or injury and arrived here in the fifth century BC, over the course of days and weeks you would have progressively ascended each of the four terraces that scale these grounds, each level dedicated to a different stage of diagnosis, counseling, and healing. Apart from simple relaxation, your treatment might have included bathing in large pools, being massaged with perfumes, oils, and ointments, following a regimen of mental and physical exercise, receiving diet counseling, herbs and other oral drugs, and commiseration with the ancient spirits.

Oh, and one more thing. If you happened to check in sometime between 490 and 377 BC, you might have received one more benefit: a visit from the world's "first" physician, a man not only credited with inventing the practice of medicine, but whose insights have remained influential for well over two thousand years.

* * *

Most of us have a distinct yet vague impression of who Hippocrates was. The phrase "Father of Medicine" often (and accurately) jumps to mind. And of course, there is the Hippocratic Oath, which we know has something to do with doctors behaving nicely. On the other hand, it should be noted that Hippocrates bears no connection to the similar-sounding "hypocrisy." Though also Greek in origin, hypocrisy is from *hypokrisis*, which means "playing a part" or, as commonly used today, someone who is a phony.

Which Hippocrates certainly was not.

Who was Hippocrates, and how did he earn the mantle of "Father of Medicine" as well as credit for the "Invention of Medicine"?

Perhaps one measure of an individual's greatness is when the question is not so much how their "breakthrough" compares to another, but rather, *which* of their many breakthroughs one should select to make the comparison. For Hippocrates, the list is substantial and includes being the first physician to

- Recognize that diseases have natural causes, rather than their arising from supernatural or evil forces
- Invent "clinical medicine" and the "doctor-patient relationship"
- Create an oath of conduct that has remained influential for 2,500 years

- Elevate the practice of medicine to an honored profession, rather than a conventional trade like plumbing or roof repair
- Achieve many other medical breakthroughs, including recognizing that thoughts and emotions arise in the brain rather than the heart

And yet…

* * *

Some time around 440 BC a young physician eager for knowledge crossed the narrow body of water separating his island home from what we know today as southwestern Turkey. Reaching land, he made his way 50 miles north to an area known as Ionia. Entering the city of Miletus, he met with the well-known philosopher Anaxagoras. Famed for introducing philosophy to the Athenians, Anaxagoras was also the first person to recognize that the moon's brightness is due to reflected light from the sun. The ensuing conversation must have been interesting. On the one hand, Hippocrates was a reputed descendent of Asklepius, the god of healing and son of Apollo. On the other hand, Anaxagoras was likely unimpressed by religious tradition: In 450 BC he had been imprisoned for insisting that the sun was not a god. While this outrageous claim may have raised the hackles of any other healer from Kos, more likely it set a twinkle in the eye of young Hippocrates. And an invitation to sit down for a chat…

* * *

And yet among the many "firsts" commonly attributed to Hippocrates, one breakthrough at the core of his teachings is often forgotten or overlooked today. Perhaps this lapse is due to its paradoxical nature, the fact that it both opposes yet resonates with the way medicine is often practiced today. What was this additional breakthrough? Before answering, we need to learn more about this man and his place in history.

The making of the man: 19 generations of healers and 3 first-rate legends

In today's high-tech world of CAT, MRI, PET, SPECT, and other cryptic visions, of the increasing specialization and molecularization of medicine, of all manner of pharmacopeia from the fitful to the fatal, we put a certain amount of trust in the rituals of modern medicine. We are comforted by hospital rooms where patients are anchored to their sanitized beds by the wires and tubes of modern technology. If for some reason then, you were to succumb to an illness and wake up in the fifth century BC in a dim, oil-lamp-lit chamber to the sound of a priest moaning incantations over your hurting body, chances are you would be overcome by a distinct lack of confidence, if not terror.

Hippocrates may well have felt the same.

Yet, born on Kos in 460 BC, this was the world in which he was raised. Like many doctors today, Hippocrates came from a line of physicians who had been practicing "medicine" for generations. For starters, he was trained in medicine by his father, Heracleides, his grandfather, and other famous teachers of the time. But this is being too modest. In fact, his family also claimed that the tradition of medicine had been in their lineage for no less than 19 generations, dating back to Asklepieios, the demi-god of healing. Deities aside, Hippocrates' early view of medicine was probably influenced by a long, long ancestry of religious healers and priests.

If you're thinking that claiming to be the nineteenth-generation descendent of the god of healing on your medical school application might strain the limits of credulity—or, conversely, that it might be just the edge you need for acceptance—several caveats are in order. First, surprisingly few undisputed details are known about the life of Hippocrates. Although a large body of writings attributed to Hippocrates have survived—some 60 works collectively known as the *Corpus Hippocraticum*, or simply Hippocratic Corpus—there is considerable debate as to which are genuine works of Hippocrates versus the embellishments of the many admirers who expanded on his school of thought decades and even centuries after his death. Nevertheless, by comparing and analyzing the documents, historians have patched together a reasonably credible account of Hippocrates and his accomplishments.

* * *

To be honest, three of the most colorful stories about Hippocrates are probably rooted as much in legend as they are truth. But even if only partly true, they provide insight into the man Hippocrates may well have been, a man whose reputation was sufficiently formidable to spread beyond his own small island to the distant lands of his own enemies.

The first and perhaps best known story is set in 430 BC during the Peloponnesian War. Shortly after being destroyed by the Spartans, a plague broke out in the city of Athens. Hippocrates and his followers traveled to Athens to help. Observing that the only people not affected by the plaque were iron smiths, Hippocrates made an astute deduction: Their resistance must somehow be related to the dry, hot atmosphere in which they worked. He promptly wrote up his prescription. The citizens of Athens were to light fires in every home to dry the atmosphere, to burn corpses, and to boil all water before consumption. The plague retreated, and Athens was saved.

The second story is often cited to highlight Hippocrates' remarkable diagnostic skills, which ranged from the physical to the psychiatric. Shortly after the Athenian plague, King Perdiccas of Macedonia, aware of Hippocrates' growing reputation, requested the physician's help when no other doctor could diagnose his vexing symptoms. Hippocrates agreed and traveled to Macedonia to see the king. During the examination, Perdiccas blushed whenever a beautiful girl named Phila—who was his father's concubine—was nearby. Hippocrates took note. Upon further inquiry, he learned that Perdiccas had grown up with Phila and dreamed of one day marrying her. This dream was shattered when his father took the girl as his concubine. However, the recent death of his father reawakened Perdiccas' conflicted feelings of love for Phila, causing him to fall ill. After subsequent counseling by Hippocrates, the king was cured.

The third story, a testament to Hippocrates' loyalty, took place when Greece was at war with Persia. By this time, Hippocrates' reputation was so great that Artaxerxes, the enemy king of Persia, requested that Hippocrates travel to Persia to save its citizens from a plague. Despite the king's offer of gifts and wealth "equal to his own," Hippocrates politely declined. Although sympathetic, it was against his scruples to assist the enemy of his country. The king gracefully responded with a vow to destroy the island of Kos—a threat that was put to rest, figuratively and literally, when the king suffered a stroke and died.

Legends aside, our investigation into the invention of medicine may be better served by looking at the achievements of Hippocrates as documented in the more scholarly writings of the Corpus. While historians continue to debate the authenticity of even these documents, said caveats having been acknowledged, we can venture into the territory where Hippocrates' "Invention of Medicine" can be attributed to six major milestones.

And yet...

* * *

While there is no record of the conversation between Hippocrates and Anaxagoras in that ancient city of Miletus, it's not hard to imagine that the young physician was beginning to question the medical tradition of his own family, with its lineage of demi-gods, superstitions, and priest healers. It wasn't that Hippocrates completely rejected their theocratic approach; he simply felt that in medicine and health, other truths were to be found. Thus, the reputation of Anaxagoras and his philosophy, which had reached even the small island of Kos, brought Hippocrates here to question and learn. Settling down in the shade of a tree outside the city, Hippocrates proffered a simple invitation. "You know of my background and tradition, Anaxagoras. Now tell me of yours..."

Milestone #1
Getting real: diseases have natural causes

> "[Epilepsy] appears to me to be nowise more divine nor more sacred than other diseases... Men regard its cause as divine from ignorance and wonder."
>
> —*Hippocrates,* On the Sacred Disease, *420-350 BC*

Until the time of Hippocrates, the most commonly accepted explanation for the cause of essentially all illness was refreshingly simple: Punishment. Having been found guilty of some misbehavior or moral failure, the gods or evil spirits exacted their justice through sickness. Your redemption, or "treatment" as we call it today, might include a visit to a nearby temple of

Asklepieios, where the local priests attempted to cure your malady with incantations, prayers, or sacrifice.

At some point early in his career, Hippocrates changed the rules. Distancing himself from the Asklepian priests and their theocratic approach to healing, Hippocrates insisted that diseases were caused by natural forces and not the gods. No statement better summarizes his view than the frequently quoted passage from one of the books attributed to Hippocrates, *On the Sacred Disease*. The title of this book—the first to be written about epilepsy—references the belief at the time that seizures were caused by the "sacred" hand of a displeased god.

Hippocrates begged to differ:

> "It appears to me to be nowise more divine nor more sacred than other diseases, but has a natural cause like other affections. Men regard its nature and cause as divine from ignorance and wonder, because it is not at all like other diseases. And this notion of its divinity is kept up by their inability to comprehend it... Those who first referred this disease to the gods appear to me to have been just such persons as the conjurers and charlatans... Such persons, then, using divinity as a pretext and screen of their own inability to afford any assistance, have given out that the disease is sacred..."

In this and similar writings, we hear in the voice of Hippocrates not only the adamancy of his view that disease arises from natural causes, but the exasperation, if not contempt, he holds for the "charlatans" who would claim otherwise. Thus with such statements, and nothing more than his own mortal powers, Hippocrates wrestled disease from the supernatural and placed it squarely in the world of the rational and natural.